Advanced Radio Frequency Identification Design and Applications

Advanced Radio Frequency Identification Design and Applications

Editor

Dan Hallmen

scitus academics

Advanced Radio Frequency Identification Design and Applications

Edited by **Dan Hallmen**

ISBN: 978-1-68117-276-7
Library of Congress Control Number: 2016937448

SCITUS Academics LLC,
www.scitusacademics.com
Box No. 4766, 616 Corporate Way,
Suite 2, Valley Cottage,
NY 10989

Notice

Printed in United States of America on Acid Free Paper

Preface

Radio frequency identification (RFID) technology uses radio wave communication to exchange data between the reader and the electronic tag attached to an object for tracking and identification. RFID is an automatic method for identifying and collecting data. A radio frequency identification system uses tags or labels attached to the objects to be identified. Two radio receivers called readers send a signal to the label and read its answer. RFID tags with battery support may be active or passive. An active tag has an on-board battery and periodically transmits its ID signal. A passive assisted battery is equipped with a small battery and activated when an RFID reader is available. A passive tag is cheaper and smaller, because it has no battery, but uses the radio energy of the reader. However, it must be illuminated at a power level about a thousand times more powerful than a passive tag for signal transmission. This makes a difference in interference and exposure to radiation. The tags contain electronically stored information. Passive tags gather energy from a nearby RFID reader's interrogating radio waves. Active tags have a local power source like a battery and can operate hundreds of meters away from the RFID reader. Unlike a barcode, the tag does not have to be in the viewing line of the reader, so it can be integrated into the tracked object. RFID tags are used in many industries, such as an RFID tag attached to an automobile during production to track its progress through the assembly line; RFID tagged pharmaceuticals can be tracked through warehouses; and the implantation of RFID microchips in animals and pets enables positive animal identification.

This book highlights the progress in RFID tag antenna and ASIC design, new chipless RFID tag design, security protocol improvements and some new RFID applications.

Table of Contents

1

CHAPTER

RADIO FREQUENCY BACKGROUND

Tales Cleber Pimenta[1], Paulo C. Crepaldi[1] and Luis H. C. Ferreira[1]

[1] *Universidade Federal de Itajuba,, Brazil*

1. INTRODUCTION

Design considerations for the traditional low frequency circuits and the RF circuits are quite different. In low frequency design, the maximum signal transfer occurs when the source presents low impedance while the load presents high impedance. A typical example is a buffer, where the input impedance is high and the output impedance is low. As long as that requirement is fulfilled, the designer is capable of choosing arbitrary levels of impedance that best suits the circuit requirements or applications.

Therefore this chapter aims to provide background on impedance matching between source and load, with or without a transmission line. The analysis can be conducted by using Smith Charts and S-Parameters, which are also presented in this chapter. The analysis in this chapter is oriented to RFID applications whereas other books provide general analysis.

During RF design, the impedances should be matched for maximum signal transfer. Additionally, when the circuits are connected using transmission lines, they should match also the standard values of the transmission lines. At very low frequencies, transmission lines can be thought as just a wire. Nevertheless, at high frequencies, the signal wavelength is comparable to or smaller than the length of the transmissions line and power can be seen as traveling waves. As a matter of fact, even a conductor can be thought as a transmission line in a high frequency circuit.

Most RF equipments and coaxial cables use the standard impedances of 50 or 75 Ω. The value of 75 Ω is used, as an example, in cable TV equipment, since this value provides the minimum losses, as it is desired in transmitting the signal over long distances. In fact, the value of impedance for minimum loss should be 77 Ω, but it was rounded to 75 Ω by convenience.

The value of 50 Ω corresponds to a reasonable compromise, the average, between the minimum loss of a 77 Ω and the maximum power handling capability given of 30 Ω.

2. TRANSMISSION LINE

Fig. 1 shows the lumped component model of a real (lossy) transmission line. The segment indicated corresponds to an infinitesimal segment of the transmission line. The characteristic impedance Z_0 of this line can be found to be [1]:

$$Z_0 = \sqrt{\frac{Z}{Y}} = \sqrt{\frac{R + j\omega L}{G + j\omega C}} \quad (1)$$

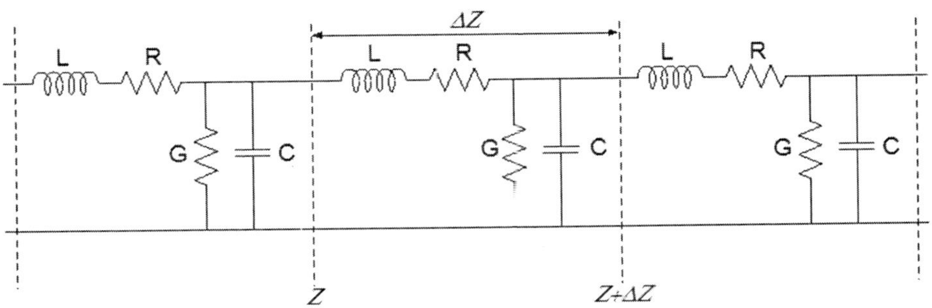

Figure 1. Lumped component model of a transmission line.

As can be observed, the characteristic impedance Z_0 is dependent on the frequency. Nevertheless, if the resistive terms R and G are negligible, the expression of the characteristic impedance Z_0 can be simplified to:

$$Z_0 = \sqrt{\frac{L}{C}} \quad (2)$$

If the value of RC is equal to GL, expression (1) yields the same value of expression (2). In other words, choosing the L/R time constant of the series impedance equals to the C/G time constant, a lossy line will behave as a lossless line, so that its characteristic impedance will be independent of the frequency[1].

2.1. Reflection coefficient

If a transmission line is terminated by an impedance Z_0, then a signal traveling down the line with a ratio of voltage to current equal to Z_0 will maintain its ratio upon encountering the load and there will be no reflections. On the other hand, when the load is different of Z_0, then it imposes its own particular ratio of voltage to current, and the only way to reconcile the conflict is by having some of the signal reflected back towards the source. In order to distinguish the incident and the reflected signals, the subscripts *i* and *r*, respectively, will be used.

The incident signal is given by:

$$Z_0 = \frac{E_i}{I_i} \tag{3}$$

At the load end, the mismatch in impedances gives rise to a reflected signal. Since the system is still linear, the total voltage at any point in the system is the sum of incident and reflected voltages. The net current is superposition of incident and reflected currents. However, since the currents are traveling in opposite directions, the net current is the difference between them. Therefore, the load impedance is given by:

$$Z_L = \frac{E_i + E_r}{I_i - I_r} \tag{4}$$

Expression (4) can be rewritten to express Z_L as function of Z_0 as:

$$Z_L = \frac{E_i + E_r}{I_i - I_r} = \frac{E_i}{I_i}\left[\frac{1 + E_r / E_i}{1 - E_r / E_i}\right] = Z_0\left[\frac{1 + E_r / E_i}{1 - E_r / E_i}\right] \tag{5}$$

The ratio of reflected to incident quantities at the load end of the line is called Reflection Coefficient Γ_L. Therefore, expression (5) can be rewritten as:

$$Z_L = \frac{E_i + E_r}{I_i - I_r} = Z_0 \left[\frac{1 + \Gamma_L}{1 - \Gamma_L} \right] \quad (6)$$

Solving for Γ_L yields:

$$\Gamma_L = \frac{Z_L - Z_0}{Z_L + Z_0} \quad (7)$$

As can be observed from expression (7), if the impedances of the load and the line are equal, there will be no reflection. If the line is terminated in either a short or an open circuit, the reflection will be maximum, with a magnitude of 1 [1].

Therefore, if a transmission line is terminated by its characteristic impedance there will be no reflection since all the transmitted power is absorbed by the load and the energy flows in just one direction.

When the line is terminated by a short circuit a reflected wave is sent back to the source since the short can not sustain any voltage, and therefore dissipates zero power. The incident and the reflected voltage waves are of the same magnitude. They are 180 out of phase at the load and they travel in opposite directions.

If the line is terminated by an open circuit a reflected wave is sent back to the source since the open can not sustain any current, and therefore dissipates zero power. The incident and the reflected current waves are of the same magnitude and travel in opposite directions. The current waves are 180 out of phase at the load, but the incident and reflected voltage waves are in phase [1].

If the line is terminated by an impedance different of the short, open and characteristic impedance, part of the signal will be absorbed by the load and part will be reflected back. The amount of reflected signal is given by expression (7).

3. SMITH CHART

The reflection coefficient Γ_L of expression (7) was obtained from expression (6). By the same way, solving for Z_L in expression (7) yields Γ_L, thus forming a mapping of one complex number into another. The relationship between these two complex numbers forms a bilinear transformation, which means that knowing one is equivalent to knowing the other.

Since Z_L can have any value and $|\Gamma_L|$ cannot exceed unity for passive loads, it is therefore much more convenient plotting Γ_L than plotting Z_L.

The reflection coefficient can become even more convenient by normalizing it to Z_0, as:

$$\Gamma = \frac{\frac{Z_L}{Z_0} - 1}{\frac{Z_L}{Z_0} + 1} = \frac{Z - 1}{Z + 1} \tag{8}$$

On the same way, normalizing (6) results in:

$$Z = \frac{1 + \Gamma}{1 - \Gamma} \tag{9}$$

Considering the normalized real and imaginary parts of both Γ and Z then:

$$Z = R + jX = \frac{1 + \Gamma}{1 - \Gamma} = \frac{1 + \Gamma_r + j\Gamma_i}{1 - \Gamma_r - j\Gamma_i} \tag{10}$$

After some algebraic manipulation (using conjugate), the real and imaginary parts are of Z are:

$$R = \frac{1 - \Gamma_r^2 - \Gamma_i^2}{1 + \Gamma_r^2 - 2\Gamma_r + \Gamma_i^2} \tag{11}$$

$$X = \frac{2\Gamma_i}{1 + \Gamma_r^2 - 2\Gamma_r + \Gamma_i^2} \tag{12}$$

Expression (11) can be manipulated as:

$$R + R\Gamma_r^2 - 2R\Gamma_r + R\Gamma_i^2 = 1 - \Gamma_r^2 - \Gamma_i^2$$

$$R\Gamma_r^2 - 2R\Gamma_r + R\Gamma_i^2 + \Gamma_r^2 + \Gamma_i^2 = 1 - R$$

$$R\Gamma_r^2 - 2R\Gamma_r + R\Gamma_i^2 + \Gamma_r^2 + \Gamma_i^2 = (1 - R)\frac{(1 + R)}{(1 + R)}$$

$$R\Gamma_r^2 - 2R\Gamma_r + R\Gamma_i^2 + \Gamma_r^2 + \Gamma_i^2 = \frac{1}{(1+R)} - \frac{R^2}{(1+R)}$$

$$\Gamma_r^2(1+R) - 2R\Gamma_r + \frac{R^2}{(1+R)} + \Gamma_i^2(1+R) = \frac{1}{(1+R)}$$

$$\Gamma_r^2 - 2\Gamma_r\frac{R}{(R+1)} + \frac{R^2}{(1+R)^2} + \Gamma_i^2 = \frac{1}{(1+R)^2}$$

$$\left(\Gamma_r - \frac{R}{R+1}\right)^2 + \Gamma_i^2 = \left(\frac{1}{1+R}\right)^2 \tag{13}$$

Similarly, expression (12) into:

$$(\Gamma_r - 1)^2 + \left(\Gamma_i - \frac{1}{X}\right)^2 = \frac{1}{X^2} \tag{14}$$

When the two parametric equations (13) and (14) are drawn on a complex coordinate, they build the Smithchart. Equation (13) forms resistance circles, and equation (14) generates reactance circles, as shown in Fig. 2 and Fig. 3, respectively. The resulting Smithchart is illustrated in Fig. 4.

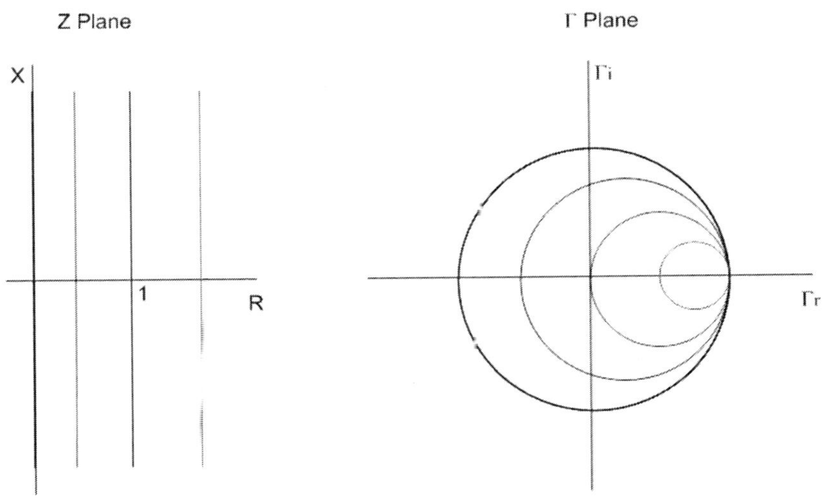

Figure 2. Mapping of Z-plane into Γ-plane for constant resistances.

As can be verified from expression (13), the imaginary axis in the Z-plane (resistance equals 0) is mapped as a unity circles into Γ-plane. The other lines of constant resistance in the Z-plane are also mapped as circles, but of different diameter in the Γ-plane. Nevertheless, they are all tangent at the point $\Gamma=1$, as shown in Fig. 2. The larger the value of the resistance, the smaller becomes the circle [1, 2].

As can be verified from expression (14), lines of constant reactance are perpendicular to lines of constant resistance in the Z-plane. This same orthogonality is preserved in the Γ-plane by having arcs perpendicular to the constant resistance lines, as indicated in Fig. 3.

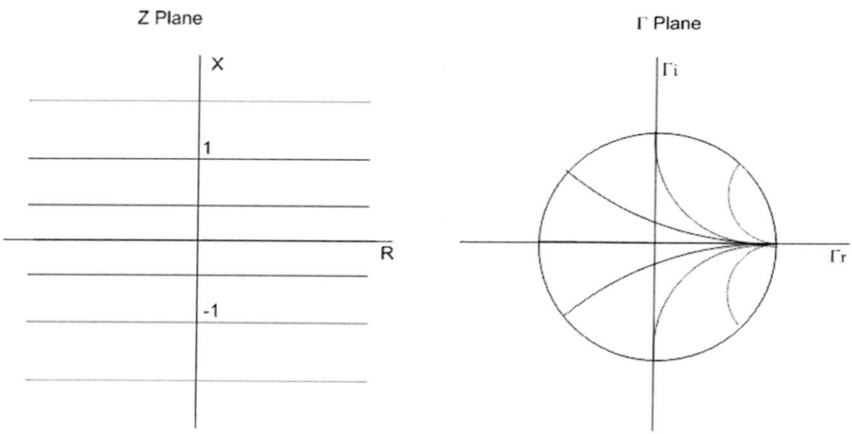

Figure 3. Mapping of Z-plane into Γ-plane for constant reactances.

The Smith, as shown in Fig. 4, is just the plotting of both constant resistance and constant reactance, but without the presence of the Γ axes. The center of the Smith chart corresponds to zero reflection (Z_L equals Z_0)[1].

The upper half of the Smith chart corresponds to the upper half part of the Z-plane, and therefore presents inductive loads. On the same way, the bottom half of the Smith chart corresponds to the bottom half part of the Z-plane, thus representing capacitive loads. Obviously, the Re axis of the Smith chart represents purely resistive loads.

Although the Smith chart presents many interesting and useful properties, they will no be presented here due to the focus of this material.

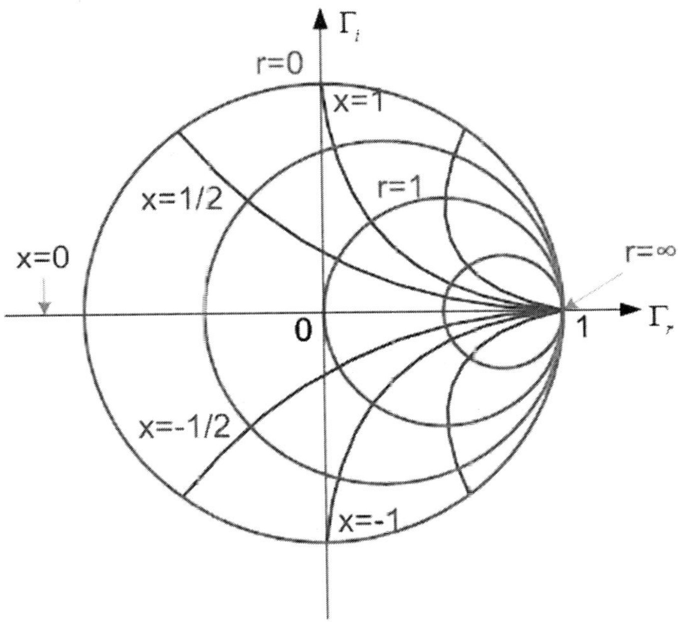

Figure 4. Smith chart.

2.1. Admittance chart

The Smithchart is built by considering impedance (resistance and reactance). Once the Smithchart is developed, a similar approach can be used for admittance analysis. The concept that admittance is the inverse of impedance is very important for parallel circuit synthesis. Adding new elements in series can be resolved easily by adding the impedance values. However, summing elements in parallel can be cumbersome in terms of impedance. Thus, admittance is often considered for parallel elements.

By definition, admittance is expressed as:

$$Y = \frac{1}{Z} = G + jB \tag{15}$$

where G is conductance and B is susceptance of the element. The reflection coefficient Γ can be expressed in terms of normalized admittance as:

$$\Gamma = \Gamma_r + j\Gamma_i = \frac{Z_L - Z_0}{Z_L + Z_0} = \frac{\frac{1}{Y_L} - \frac{1}{Y_0}}{\frac{1}{Y_L} + \frac{1}{Y_0}} = \frac{Y_0 - Y_L}{Y_0 + Y_L} = \frac{1-y}{1+y} = \frac{1-g-jb}{1+g+jb} \tag{16}$$

The admittance reflection coefficient *Γ(y)* is given by definition as:

$$\Gamma_{(Y)} = \frac{Y_L - Y_0}{Y_L + Y_0} \tag{17}$$

As can be observed, the value of the admittance reflection coefficient *Γ(y)* is equal to –*Γ*, the reflection coefficient in terms of impedance. Therefore, once *Γ* is known, *Γ(y)* can be located a point at the same distance from the origin. Rotating the admittance point by 180 around the origin achieves the same result.

The admittance Smithchart can be obtained using the same procedure used to construct the impedance Smithchart. The normalized real and imaginary parts of *Γ* and *Y* can be given as:

$$Y = G + jB = \frac{1 - \Gamma_r^2 - \Gamma_i^2 - j2\Gamma_i}{1 + \Gamma_r^2 + 2\Gamma_i + \Gamma_i^2} \tag{18}$$

After some algebraic manipulation (using conjugate), the real and imaginary parts are of *Y* are:

$$G = \frac{1 - \Gamma_r^2 - \Gamma_i^2}{1 + \Gamma_r^2 + 2\Gamma_r - \Gamma_i^2} \tag{19}$$

$$B = \frac{-2\Gamma_i}{1 + \Gamma_r^2 + 2\Gamma_r + \Gamma_i^2} \tag{20}$$

Using the same procedure presented for expressions (13) and (14), then the parametric equations of the admittance Smithchart are:

$$\left(\Gamma_r + \frac{G}{G+1}\right)^2 + \Gamma_i^2 = \left(\frac{1}{1+G}\right)^2 \tag{21}$$

$$\left(\Gamma_r + 1\right)^2 + \left(\Gamma_i + \frac{1}{B}\right)^2 = \frac{1}{B^2} \tag{22}$$

When the two parametric equations (21) and (22) are drawn on a complex coordinate, they build the Admittance Smithchart. Equation (21) forms resistance circles, and equation (22) generates reactance circles, as shown in Fig. 5.

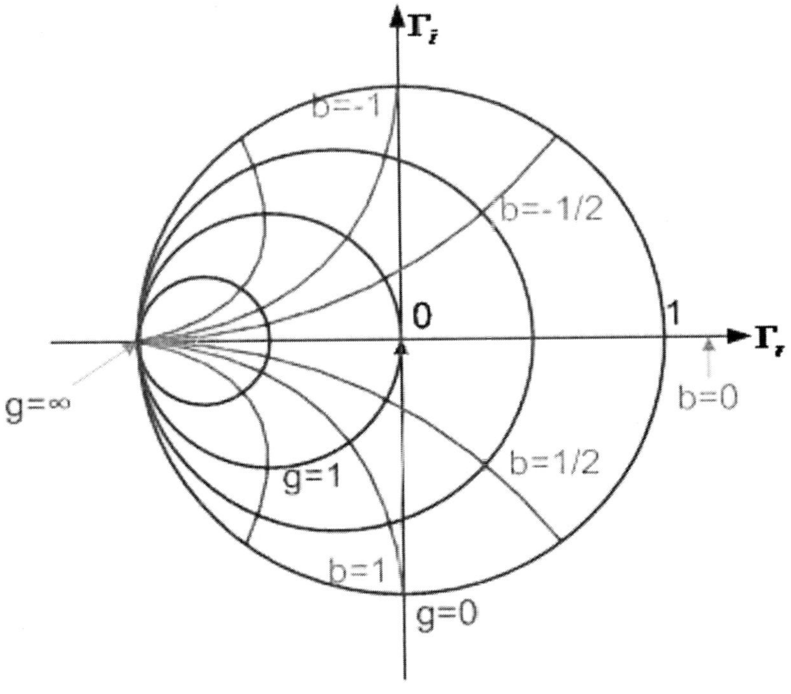

Figure 5. Admittance Smithchart.

3. S PARAMETERS

At low frequencies, linear systems can be analyzed by means of voltages and currents applied to its ports. The two port circuit shown in Fig. 5 could be analyzed from its impedance (Z-parameters), admittance (Y-parameters), or a mixture of them, which could be hybrid (H-parameters) and inverse-hybrid (G-parameters).

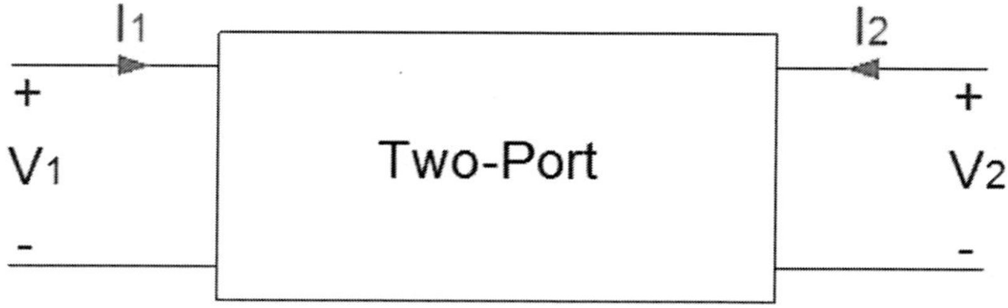

Figure 6. Two port circuit representation.

As an example, the circuit of Fig. 5 could be analyzed using H parameters whose equations are:

$$V_1 = h_{11}I_1 + h_{12}V_2$$
$$I_2 = h_{21}I_1 + h_{22}V_2 \tag{23}$$

As can be observed from the equations, the value of h_{11} (port 1 impedance) can be obtained directly from the relationship of V_1 and I_1 when V_2 is set to zero. A voltage source is set to zero by shortening its terminals. The value of h_{21} (current gain from port 1 to port 2) is obtained from the relationship of I_1 and I_2 also when V_2 is set to zero.

By the same way, h_{12} (voltage gain from port 2 to port 1) can be obtained from the relationship of V_1 and V_2 when I_1 is set to zero. A current source is set to zero by opening its terminals. Finally, h_{21} (port 2 conductance) is obtained from the relationship of V_2 and I_2 when I_1 is set to zero.

Shorting or opening terminals is feasible at low frequencies but virtually impossible at high frequencies, particularly over a broad range of frequencies. Additionally, RF circuits are very sensitive to impedances, and they may oscillate or just quit working when terminated with open or short circuits. Therefore, Z-parameters, Y-parameters, H-parameters and G-parameters are not suitable to high frequency operations.

For high frequency operations, the *scattering* parameters (or just S-parameters) are employed. The input and output variables in S-parameters are based on incident and reflected voltage waves instead of voltages and currents.

The S-parameters takes advantage of the fact that there is no reflection in a line terminated in its characteristic impedance. Therefore, it is necessary a circuit representation for S-parameters, where source and the load terminations are Z_0, as shown in Fig. 6.

The S-parameters equations are:

$$\begin{aligned} b_1 &= s_{11}a_1 + s_{12}a_2 \\ b_2 &= s_{21}a_1 - s_{22}a_2 \end{aligned} \tag{24}$$

where

$$\begin{aligned} a_1 &= E_{i1} / \sqrt{Z_0} \\ a_2 &= E_{i2} / \sqrt{Z_0} \\ b_1 &= E_{r1} / \sqrt{Z_0} \\ b_1 &= E_{r2} / \sqrt{Z_0} \end{aligned} \tag{25}$$

The normalization by $\sqrt{Z_0}$ is very convenient since the square of a and b corresponds to the power of the incident and reflected waves.

S_{11} and S_{21} can be obtained by measuring the incident, the reflected and the transmitted signals at the input when the output is terminated in Z_0. Once the output is terminated by Z_0 there is no reflection. The values of S_{11} and S_{21} are:

$$\begin{aligned} S_{11} &= \frac{b_1}{a_1} = \frac{E_{r1}}{E_{i1}} = \Gamma_1 \\ S_{21} &= \frac{b_2}{a_1} = \frac{E_{r2}}{E_{i1}} \end{aligned} \tag{26}$$

Similarly, S_{12} and S_{22} can be obtained by measuring the incident, the reflected and the transmitted signals at the output when the input is terminated in Z_0. Since the input is terminated by Z_0 there is no reflection. The values of S_{12} and S_{22} are:

$$S_{12}=\frac{b_1}{a_2}=\frac{E_{r1}}{E_{r2}}\quad S_{22}=\frac{b_2}{a_2}=\frac{E_{r2}}{E_{i2}}=\Gamma_2 \tag{27}$$

S_{11} corresponds to the input reflection coefficient, S_{21} is the input to output (direct) gain, S_{12} is the reverse transmission gain and S_{22} corresponds to the output reflection coefficient.

The magnitudes of S_{11} and S_{22} are always less than 1, where a value of zero represents a perfect matching (no reflections), and the closer to 1, the higher the reflection.

The magnitudes of S_{21} (transfer characteristic) and S_{12} (reverse) are smaller than 1 for passive circuits but can exceed 1 for active circuits (amplification). A positive value means the input and output signal are in phase and a negative indicates a phase shift.

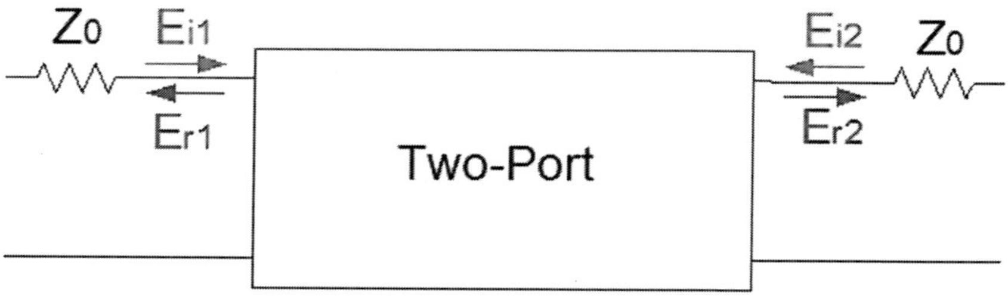

Figure 7. Two port circuit representation for S-parameters.

3.1. Measurements of S-Parameters

The circuit topology used to measure S-Parameters is given in Fig. 7.

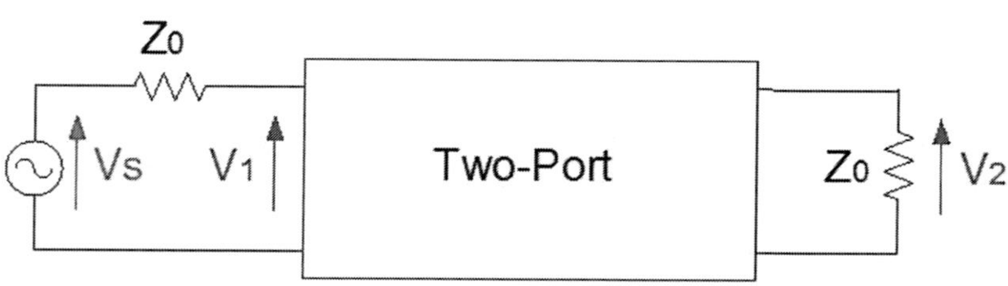

Figure 8. Circuit topology used to measure S-Parameters.

The input reflection coefficient S_{11} from expression (18) can be modified as:

$$S_{11} = \Gamma_1 = \frac{Z_L - Z_0}{Z_L - Z_0} = 2.\frac{Z_L}{Z_L - Z_0} - 1 \tag{28}$$

This expression, using the concepts of voltage division, corresponds to:

$$S_{11} = \Gamma_1 = 2.\frac{Z_L}{Z_L - Z_0} - 1 = 2.\frac{V_1}{V_S} - 1 \tag{29}$$

Here, Z_L corresponds to the input impedance of the two-port circuit. The value of the input to output gain, S_{21} is given as:

$$S_{21} = 2.\frac{V_2}{V_S} \tag{30}$$

The value of S_{22} and S_{12} can be obtained in a similar way by just exchanging input and output terminals.

3.1. S Parameters in the Smith chart

The center point of the Smith chart corresponds to the point of zero reflection, where Z_L equals Z_0. Plots of S_{11} and S_{22} on the real axis represent ohmic resistors, above the axis indicate inductive load while below the axis indicate capacitive loads.

Plots of S_{12} and S_{21} inside the Smith chart indicate that damping signal between ports, whereas plots outside the chart indicate amplification [1].

As the frequency increases, the S-Parameters plots in the Smith chart move clockwise.

Given the value of S_{11}, the circuit impedance can be found from (6), as:

$$Z_L = Z_0 \left[\frac{1 + S_{11}}{1 - S_{11}} \right] \tag{31}$$

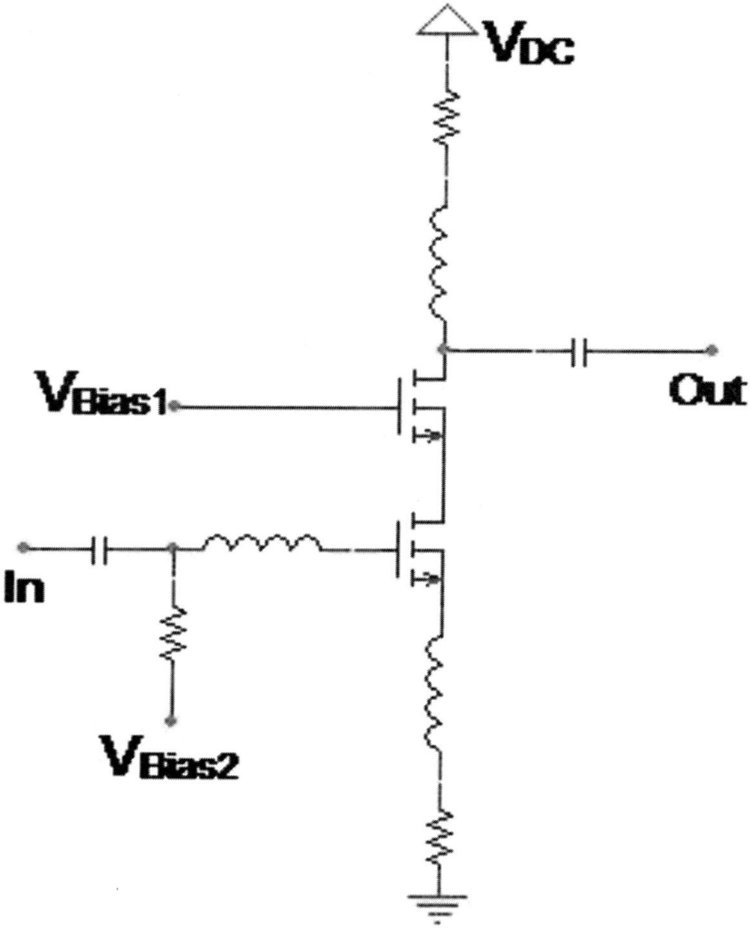

Figure 9. Cascode amplifier.

3.2. Application example

The S_{11} and S_{21} parameters can be obtained as given by expressions (21) and (22), indicated in Fig. 7, and the values are calculated as for S_{11} and S_{21}, respectively. As an application example, consider the cascode amplifier shown in Fig. 8 [2].

Fig. 10 shows its S_{11} parameter, in the Smith chart. The frequency is ranging from 4GHz to 6GHz. As the frequency increases, the plot moves clockwise. At approximately 5GHZ, the circuit presents a pure resistive impedance of approximately 50Ω (it crosses the horizontal axis). The circuit presents a capacitive behavior for frequencies bellow 5HGz and an inductive behavior for higher frequencies [1].

The same parameters could be plotted in a standard dB format, as shown in Fig. 11. The graph of Fig. 10 provides more information and insight than the graph of Fig. 11. The last one provides only the magnitude, whereas the first one provides both the imaginary and real part, so that it is possible to infer a capacitive and/or inductive behavior of the circuit, among other information.

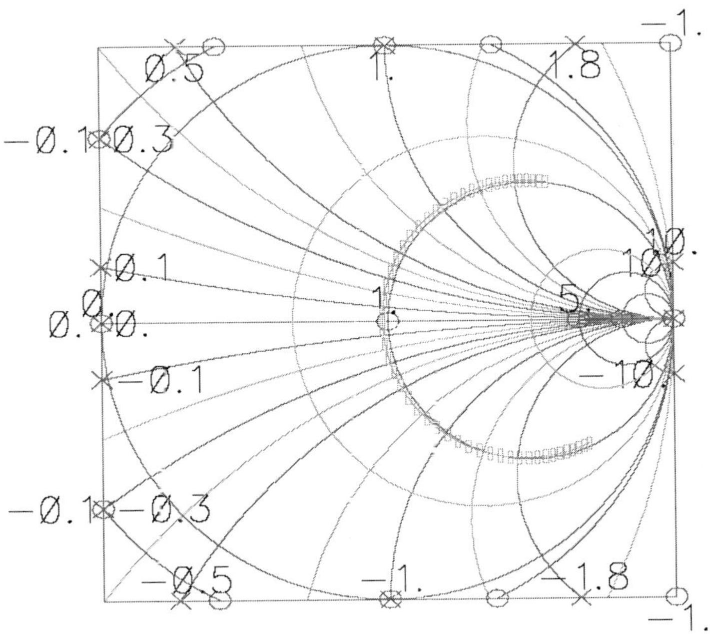

Figure 10. S_{11} parameter of the circuit from Fig. 8, in a Smith chart.

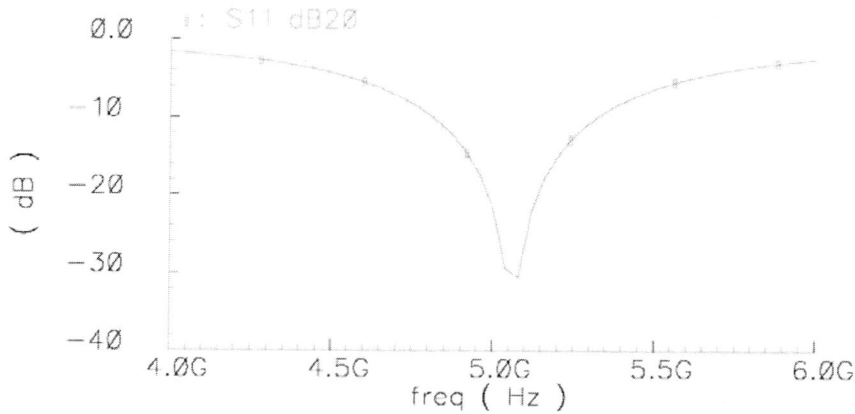

Figure 11. S_{11} parameter of the circuit from Fig. 8, in a dB chart.

Unfortunately, it is not always possible to analyze S-parameters using Smith chart. One such case is S_{11} that is usually larger than 1 for active circuits. If it is larger than one, it does not fit the Smith chart!

4. NOISE FIGURE/FACTOR

In analog circuits at low frequency, the signal-to-noise figure (SNR), defined as the ratio of signal power to the noise power, is an important and very used parameter. As an example, in a radio receiver, it indicates the quality of the demodulated signal [2-5].

Nevertheless, as the signal passes through the RF circuits, the SNR changes. This signal-to-noise degradation along the system is described by the noise factor (F), as:

$$F = \frac{SNR_i}{SNR_o} = \frac{S_i / N_i}{S_o / N_o} \tag{32}$$

where the index *i* refers to input and the index *o* refers to output.

If a system has no noise, then $SNR_o = SNR_i$ regardless of the gain, and $F=1$. This would be case of a hypothetical noiseless amplifier. Therefore, the noise figure of a real system will be always larger than 1.

Considering A as the system gain, expression (24) can be modified to:

$$F = \frac{S_i/N_i}{S_o/N_o} = \frac{S_i/N_i}{S_i.A/N_o} = \frac{N_o}{A.N_i} \tag{33}$$

which can be seen as the total output noise power over the output noise due to the input source.

The total output noise is the sum of the original noise at the input (which was amplified) and the noise added by the circuit. This can be denoted as:

$$N_o = A.N_i + N_{added} \tag{34}$$

Therefore, expression (18) can be expressed also as:

$$F = \frac{N_o}{A.N_i} = \frac{A.N_i + N_{added}}{A.N_i} = 1 + \frac{N_{added}}{A.N_i} \tag{35}$$

Again, if the circuit adds no noise, F becomes 1.

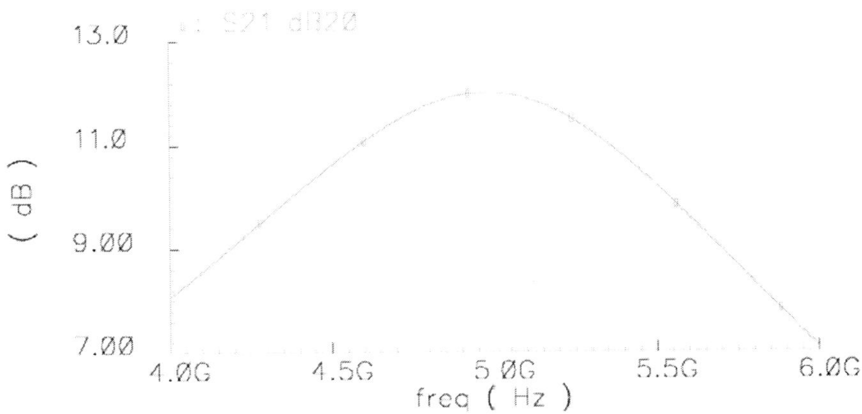

Figure 12. S_{21} parameter of the circuit Fig. 8, in a dB chart.

Another important figure of merit is the noise figure, *NF*, expressed as:

$$NF = 10.\log_{10} F \tag{36}$$

While the noise factor of a noiseless circuit is 1, the noise figure is 0dB.

4.1. Noise figure of a cascade system

Fig. 12 shows a cascade amplifying system, whose gain of each stage is A_i and the noise added by each stage is $N_{i\text{-}added}$.

The output noise due to the source is:

$$N_{o-source} = N_i.A = N_i.A_1.A_2.A_3 \tag{37}$$

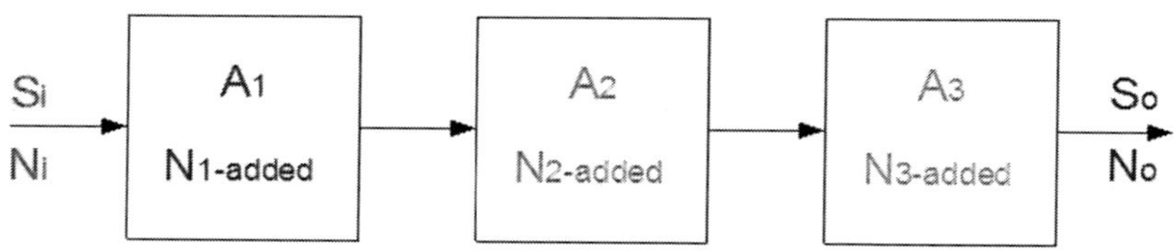

Figure 13. Cascade amplifying system.

While the total output noise is:

$$N_o = N_i.A_1.A_2.A_3 + N_{1-added}.A_2.A_3 + N_{2-added}.A_3 + N_{3-added} \tag{38}$$

which is the input noise multiplied by the gain of the three stages, plus the noise generated by the first stage and amplified by the stages 2 and 3, plus the noise generated by the second stage and amplified by the last stage, and plus the noise generated by the last stage.

Thus, combining expressions (29) and (30) into expression (25), the noise factor can be found as:

$$F = \frac{N_o}{N_{o-source}} = 1 + \frac{N_{1-added}}{N_i A_1} + \frac{N_{2-added}}{N_i A_1 A_2} + \frac{N_{3-added}}{N_i A_1 A_2 A_3} \tag{39}$$

This expression, with the aid of expression (27), can be re-written as:

$$F = F_1 + \frac{F_2 - 1}{A_1} + \frac{F_3 - 1}{A_1 A_2} \tag{40}$$

As can be observed from expression (32), the noise factor of the first stage is the most relevant to the total noise factor. That is the reason for putting most effort in the first stage in terms of noise minimization, thus requiring low noise amplifiers at the front of the system.

5. CONCLUSIONS

The basic knowledge of impedance matching between source and load, either with or without a transmission line is essential to the design of RF circuits. The analysis presented can be conducted by using Smith Charts and S-Parameters. The analysis in this chapter was oriented to RFID applications whereas other books provide general analysis.

REFERENCES

1. K. Kurokawa, 1965 Power Waves and the Scattering Matrix, IEEE Trans. Microwave Theory and Tech., 13 March 1965, 194202 , 0018-9480

2. T. H. Lee, 2004 The Design of CMOS Radio-Frequency Integrated Circuits- 2nd Edition, Cambridge University Press, 0-52183-539-9

3. J. Rogers, C. Plett, 20030 Radio Frequency Integrated Circuit Design, Artech House Inc, 1-60783-979-2

4. C. Coleman, 20040 An Introduction to Radio Frequency Engineering, Cambridge University Press, 0-52183-481-3

5. B. Razavi, 1998 RF Microelectronics, Prentice Hall, 0-13887-571-5

2
CHAPTER

TRI-BAND FRACTAL ANTENNAS FOR RFID APPLICATIONS

Abdelati Reha[1], Ahmed Oulad Said[2]

[1]Department of Telecommunications Engineering, Higher Institute of Applied Engineering (IGA), Marrakesh, Morocco; [2]Department of Electrical and Telecommunications Engineering, Royal Air Academy (ERA), Marrakesh, Morocco.

ABSTRACT

The RFID is a rapidly developing technology. It's used in many applications such as logistics, ticketing, security, employee attendance record and others. Also, fractal technology is used in many areas, and recently in antenna design because it allows making multi-band and wide-band antennas. In this paper, two tri-band fractal antennas are studied for the Radio Frequency Identification (RFID) applications using the Method of Moment (MoM). The first one is designed for the RFID readers and it operates at 3.85 GHZ, 5.80 GHZ and 8.12 GHZ. The second one is designed for the RFID Tags and it operates at 3.94 GHZ, 5.65 GHZ and 8.20 GHZ.

1. INTRODUCTION

RFID stands for Radio Frequency Identification, a term that describes any system of identification wherein an electronic device that uses radio frequency or magnetic field variations to communicate is attached to an item. The two most talked-about components of an RFID system are the tag, which is the identification device attached to the item we want to track, and the reader, which is a device that can recognize the presence of RFID tags and read the information stored on them. The reader can then inform another system about the presence of the tagged items. The system with which the reader communicates usually runs software that stands between readers and applications. This software is called RFID middleware [1].

In a typical RFID system [2], passive tags are attached to an object such as goods, vehicles, humans, animals, and shipments, while a vertical/circular polarization antenna is connected to the RFID reader. The RFID reader and tag can radio-communicate with each other using a number of different frequencies, and currently most RFID systems use unlicensed spectrum. The common frequencies used are low frequency (125 KHz), high frequency (13.56 MHz), ultra high frequency (860 - 960 MHz/2.45 GHz), and microwave frequency (3.6/3.9/5.8/ 5.9/8.2 GHZ [3]). The typical RFID readers are able to read (or detect) the tags of only a single frequency but multimode readers are becoming cheaper and popular which are capable of reading the tags of different frequencies [4].

Several calculation methods are adopted to solve the Maxwell equations and then study the performance of the antennas. Among these, two are widely used in simulation software:

1) The method of moments (MoM) is used among others in FEKO software (an abbreviation derived from the German phrase FEldberechnung bei Körpern mit beliebiger Oberfläche (Field computations involving bodies of arbitrary shape)) [5].

2) The method of finite integral (FIT) is used in the software CST Microwave Studio (Computer Simulation Technology Microwave Studio) [6].

In this work, we use the FEKO software.

2. FRACTAL ANTENNAS

The fractal antennas use fractals and self-similar models. Several studies show that we can easily make compact, multi-band and wide-band antennas, suitable for miniaturization and various applications such as RFID, WLAN (Wireless Local Area Network), WIMAX (WorldWide Interoperability for Microwave Access), WIFI (WIreless FIdelity) [7-9].

3. ANTENNAS GEOMETRY

3.1. Readers Antennas

The reader antenna is a fractal antenna 2nd iteration (Figure 1).

The antenna is fed by a Coplanar Waveguide (CPW). The antenna, the feeding point and the ground plane are on the same plane (Figure 2). Several studies have used this mode of feeding [10-13].

The antenna is designed on an FR4 substrate (relative dielectric constant = 4.3, thickness = 1.6 mm, loss tangent = 0.025). The overall dimension of the substrate is 40 × 60 mm^2.

The dimensions of the antenna are: L_1 = 60 mm, L_2 = 40 mm, L_3 = 22 mm, H = 1.6 mm, H_1 = 23 mm, H_2 = 21 mm, H_3 = 10 mm, W = 2.2 mm, S = 1 mm, G = 0.5 mm, q = 20°, a = 40° (Figure 3).

3.2. Tag Antenna

The Tag antenna is a fractal antenna 3rd iteration (Figure 4).

The antenna is fed by a Coplanar Waveguide (CPW). The antenna, the feeding point and the ground plane are on the same plane (Figure 5).

The antenna is designed on an FR4 substrate (relative dielectric constant = 4.3, thickness = 1.6 mm, loss tangent = 0.025). The overall dimension of the substrate is 30 × 90 mm^2.

The dimensions of the antenna are: L_1 = 90 mm, L_2 = 30 mm, L_3 = 10 mm, H = 1.6 mm, W = 2.4 mm, S = 1 mm, G = 1 mm (Figure 6).

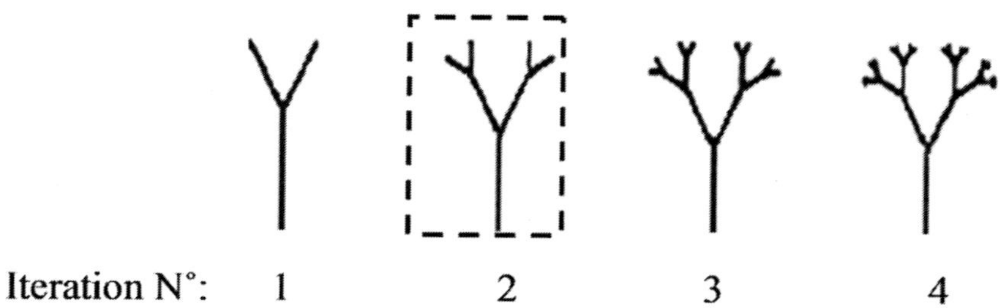

Figure 1. The first 4 iterations of the Tree-Fractal antenna.

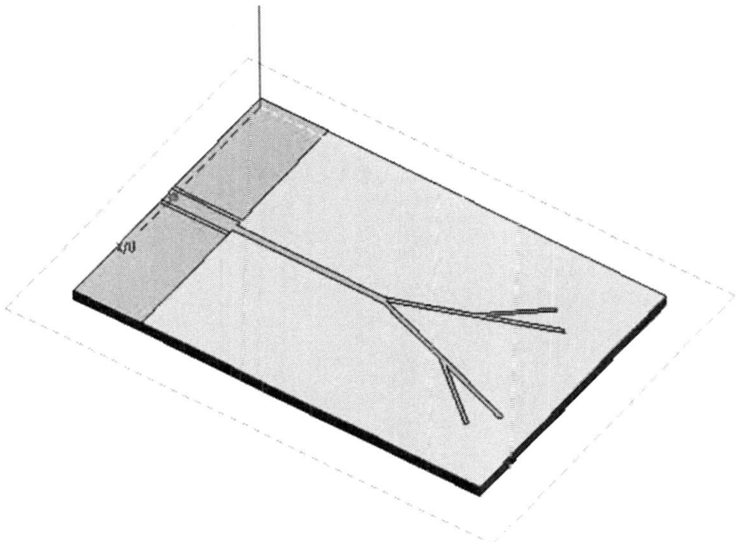

Figure 2. The Reader antenna fed by a coplanar waveguide (CPW).

4. RESULTS AND DISCUSSIONS

4.1. Reader Antenna

The analysis of the S_{11} parameter (Figure 7) shows that the Reader antenna operates at f_1 = 3.85 GHZ, f_2 = 5.8 GHZ and f_3 = 8.12 GHZ.

F_1 = 3.85 GHZ

Around this frequency, the S_{11} parameter shown by Figure 8, we note that the bandwidth is 305 MHZ (from 3.69 GHz to 4 GHZ).

We also note that the gain of this antenna is 2.6 dBi for j = 90° and j = 60° as shown by Figure 9.

F_2 = 5.8 GHZ

Around this frequency, the S_{11} parameter shown by Figure 10, we note that the bandwidth is 201 MHZ (from 5.743 GHZ to 5.944 GHZ)

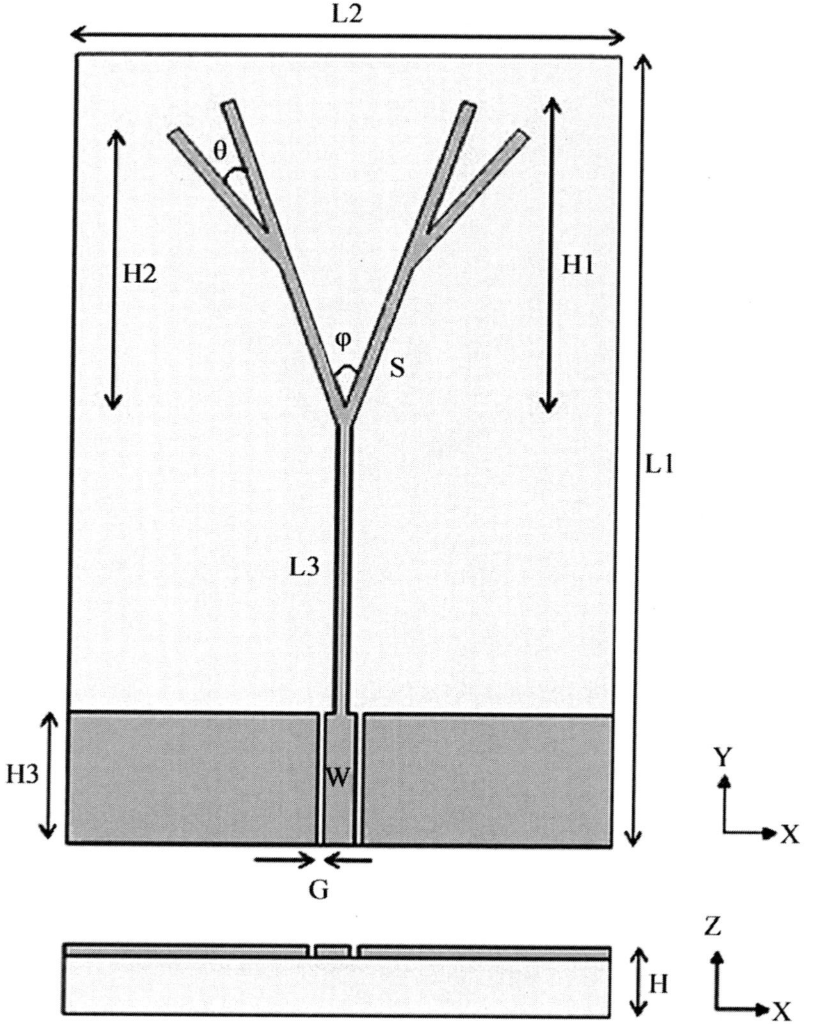

Figure 3. Dimensions of the Reader Antenna [3].

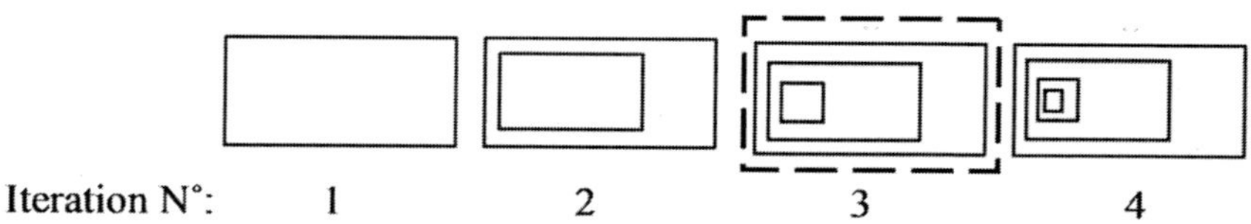

Figure 4. The first 4 iterations of the Rectangular-Fractal antenna.

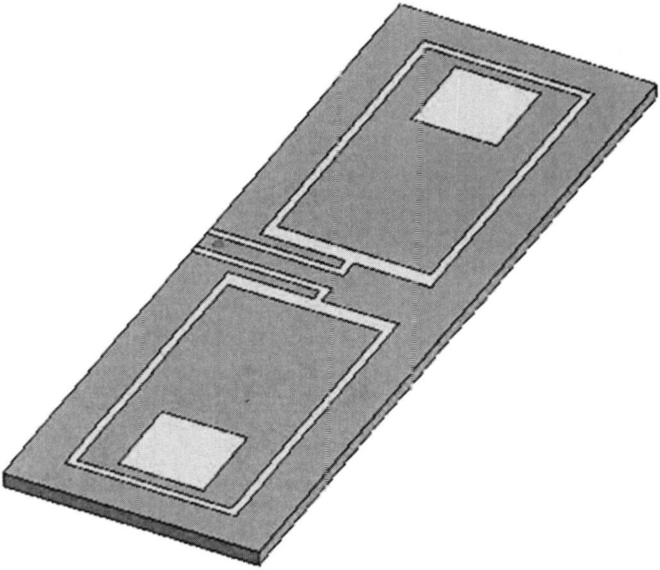

Figure 5. The Tag antenna fed by a coplanar waveguide (CPW).

We also note that the gain of this antenna is 1.7 dBi for j = 90° and j = 270° as shown by Figure 11.

F_3 = 8.12 GHZ

Around this frequency, the S_{11} parameter shown by Figure 12, we note that the bandwidth is 289 MHZ (from 7.989 GHZ to 8.278 GHZ).

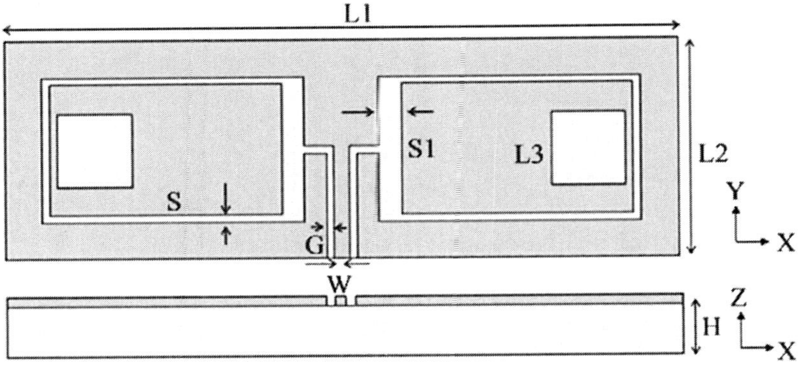

Figure 6. Dimensions of the Tag Antenna [3].

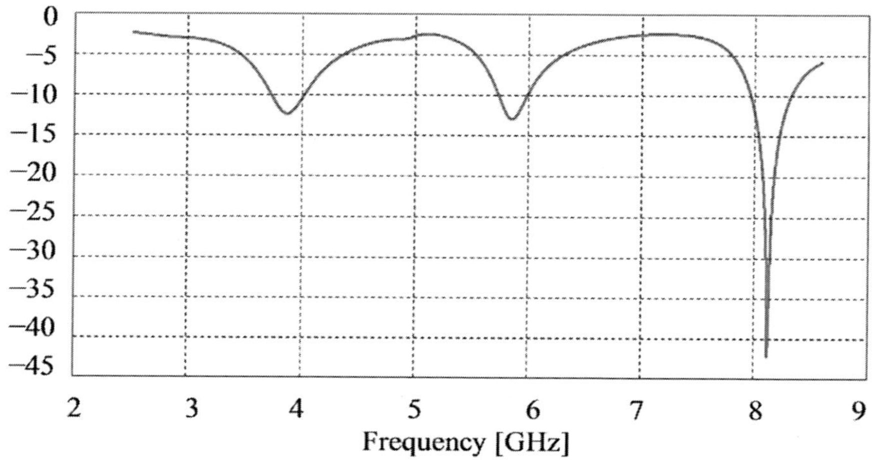

Figure 7. Simulated S_{11} versus frequency graph of the Reader antenna.

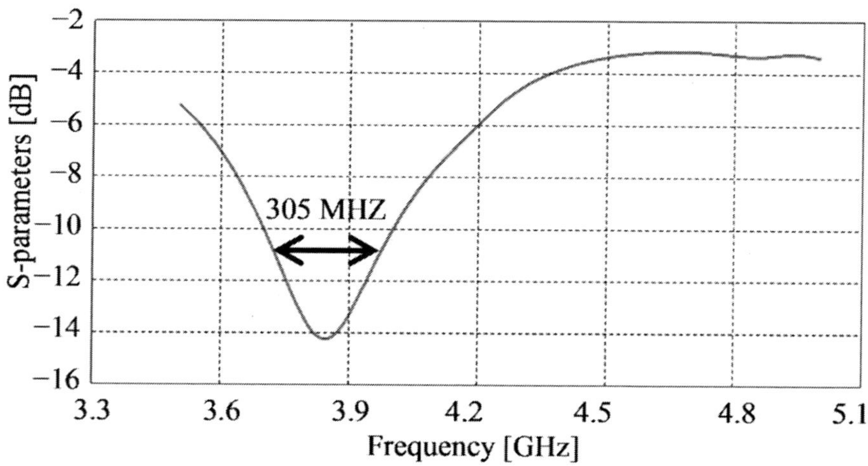

Figure 8. Evolution of S_{11} around the frequency f_1 = 3.85 GHZ.

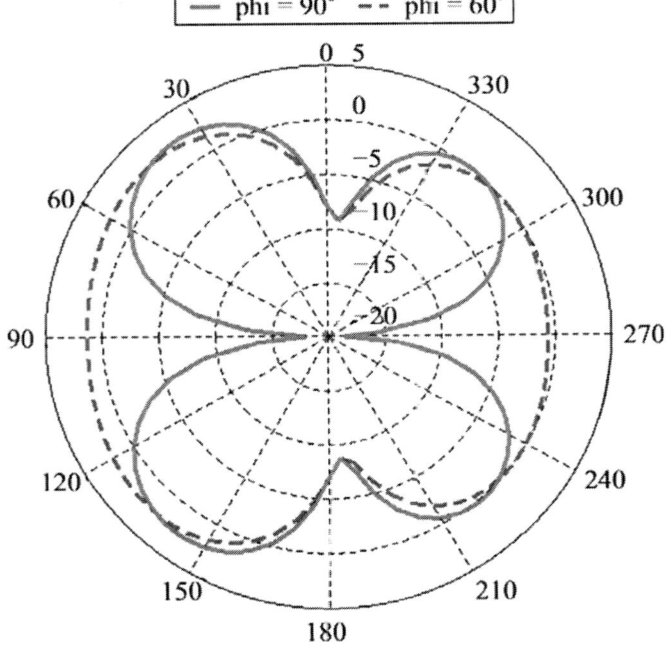

Figure 9. The gain of the antenna for f_1 = 3.85 GHZ (for j = 90° and j = 60°) – G_{max} = 2.6 dBi.

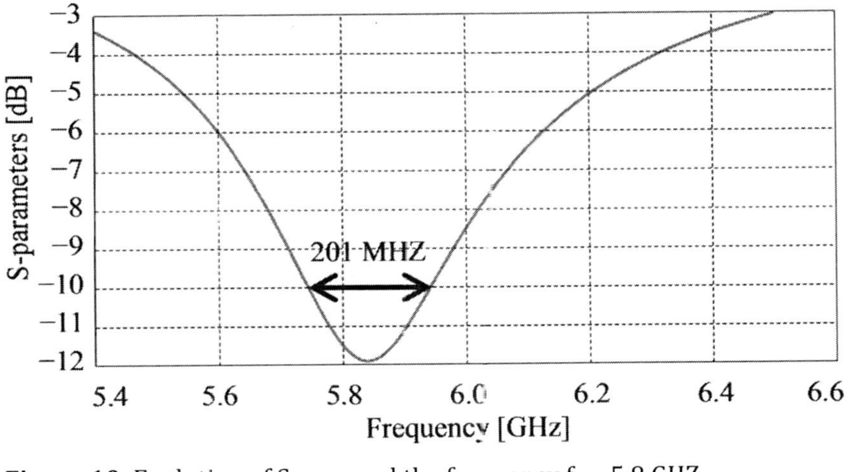

Figure 10. Evolution of S_{11} around the frequency f_2 = 5.8 GHZ.

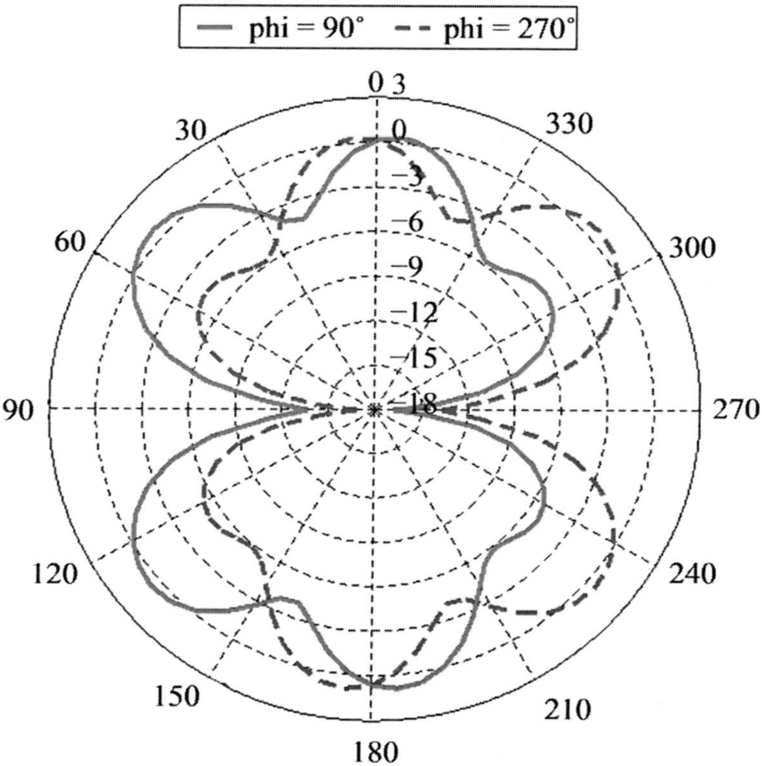

Figure 11. The gain of the antenna for f_2 = 5.8 GHZ (for j = 90° and j = 270°) – G_{max} = 1.7 dBi.

We also note that the gain of this antenna is 2.7 dBi for j = 30° and j = 90° as shown by Figure 13.

4.2. Tag Antenna

The analysis of the S_{11} parameter (Figure 14) shows that the Tag antenna operates at f_1 = 3.94 GHZ, f_2 = 5.65

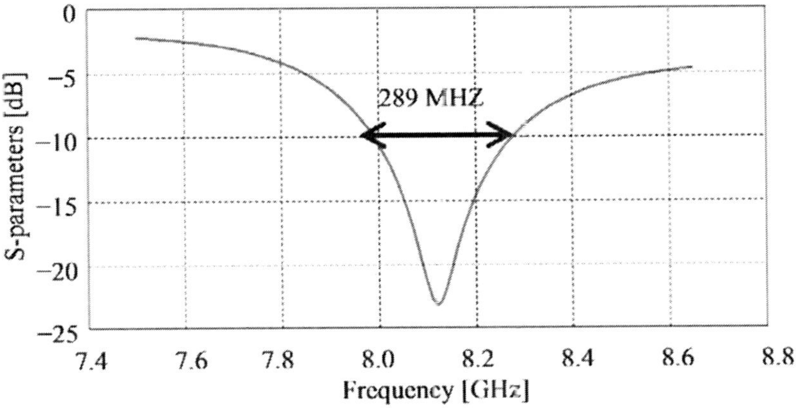

Figure 12. Evolution of S_{11} around the frequency f_3 = 8.12 GHZ.

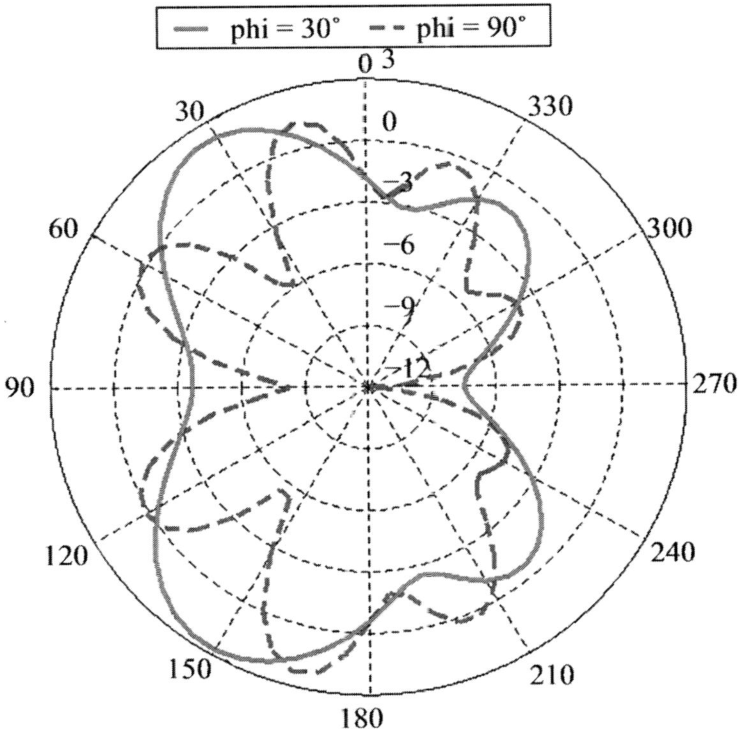

Figure 13. The gain of the antenna for f_3 = 8.12 GHZ (for j = 30° and j = 90°) – G_{max} = 2.7 dBi.

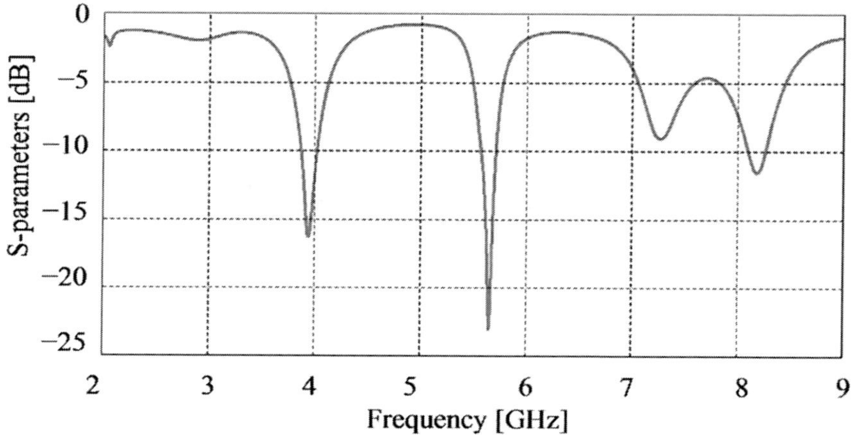

Figure 14. Simulated S_{11} versus frequency graph of the Tag antenna.

GHZ and f_3 = 8.2 GHZ.

F_1 = 3.94 GHZ

Around this frequency, the S_{11} parameter shown by Figure 15, we note that the bandwidth is 165 MHZ (from 3.869 GHZ to 4.0348 GHZ).

We also note that the gain of this antenna is 3.8 dBi for j = 0° and j = 180° as shown by Figure 16.

F_2 = 5.65 GHZ

Around this frequency, the S_{11} parameter shown by Figure 17, we note that the bandwidth is 149 MHZ (from 5.564 GHZ to 5.713 GHZ).

We also note that the gain of this antenna is 1.25 dBi for j = 0° and j = 180° as shown by Figure 18.

F_3 = 8.2 GHZ

Around this frequency, the S_{11} parameter shown by Figure 19, we note that the bandwidth is 166 MHZ (from 8.101 GHZ to 8.267 GHZ).

We also note that the gain of this antenna is 2.65 dBi for j = 0° as shown by Figure 20.

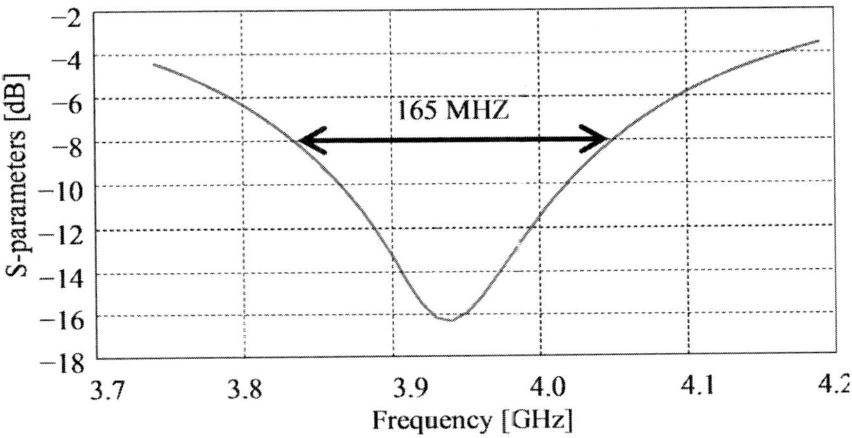

Figure 15. Evolution of S_{11} around the frequency f_1 = 3.94 GHZ.

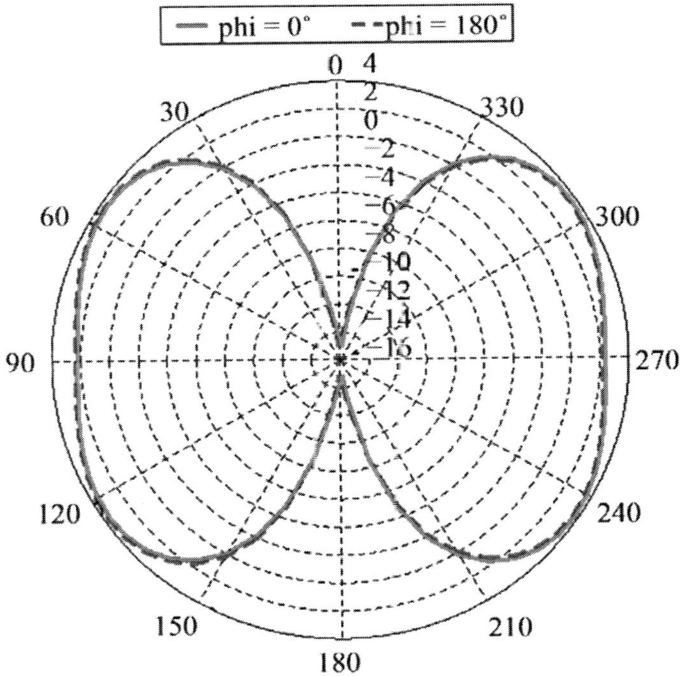

Figure 16. The gain of the antenna for f_1 = 3.94 GHZ (for j = 0° and j = 180°) – G_{max} = 3.8 dBi.

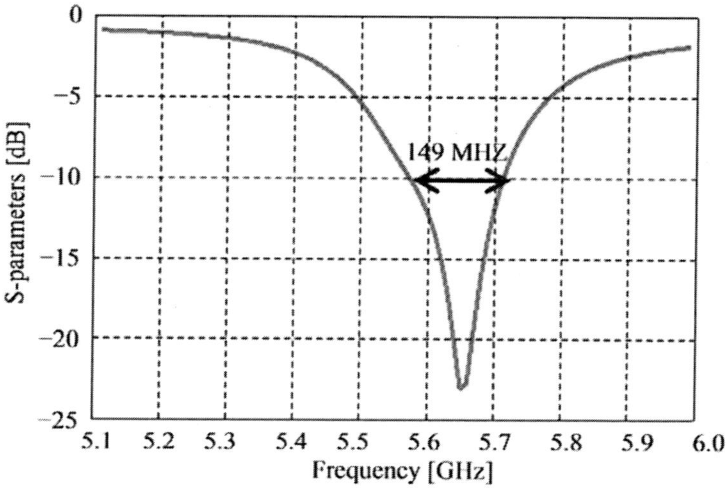

Figure 17. Evolution of S_{11} around the frequency f_2 = 5.65 GHZ.

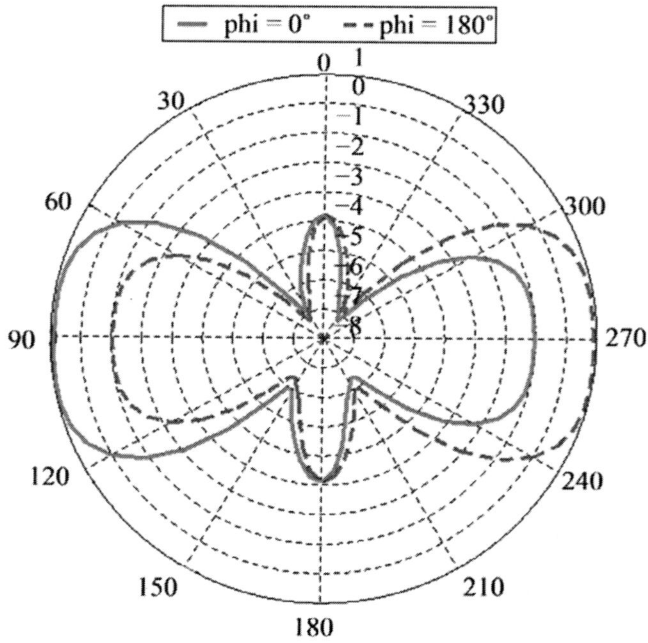

Figure 18. The gain of the antenna for f_2 = 5.65 GHZ (for j = 0° and j = 180°) – G_{max} = 1.25 dBi.

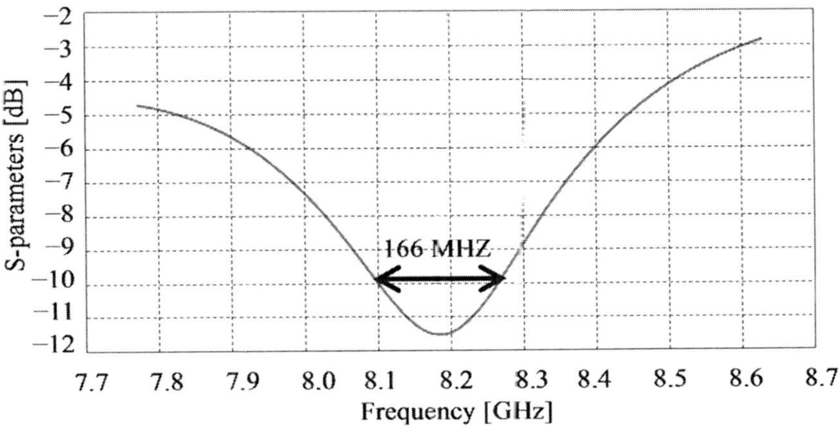

Figure 19. Evolution of S_{11} around the frequency f_3 = 8.2 GHZ.

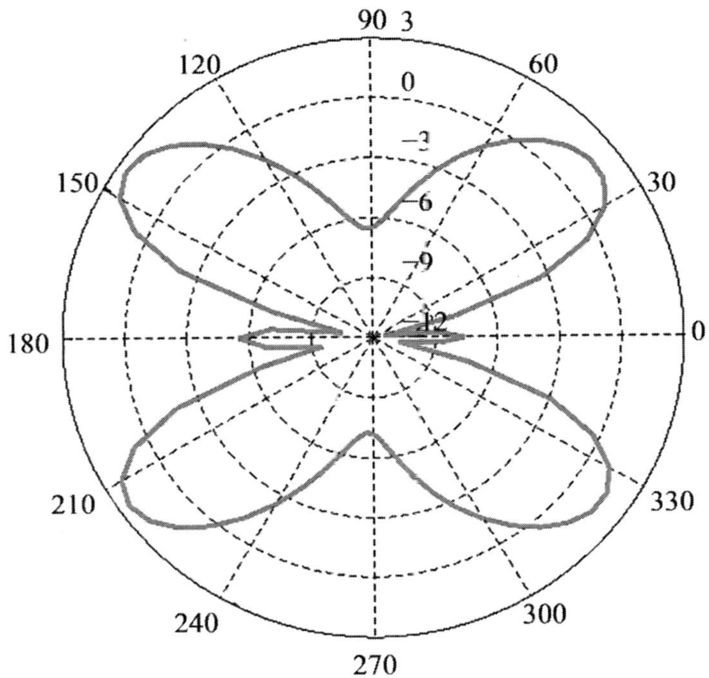

Figure 20. The gain of the antenna for f_3 = 8.2 GHZ (for j = 0°) – G_{max} = 2.65 dBi.

5. READ RANGE OF THE ANTENNAS

To calculate the read range of these antennas, we apply the First Transmission Equation (EQ1) which gives the relationship between the received power (Pr) as a function of the transmitted power (Pt), the antennas gain (Gt, Gr) and free space loss which can be expressed in terms of the wavelength (l) and the distance between the antennas(R) .

$$\frac{P_r}{P_t} = G_t \cdot G_r \left(\frac{\lambda}{4 \cdot \pi \cdot R} \right)^2 \qquad \text{(EQ1)}$$

$$R = \frac{c}{4 \cdot \pi \cdot f} \cdot \sqrt{\frac{P_t \cdot G_t \cdot G_r}{P_r}} \qquad \text{(EQ2)}$$

$$\lambda = \frac{c}{f} \qquad \text{(EQ3)}$$

For each one of the antennas, the read range calculation is done by taking a reference antenna of gain 3 dBi and minimum detectable power of 1 mW. The transmitting power is 4 W [3]. Therefore, we can build the table below (Table 1).

Table 1. Read range antennas.

frequency (GHZ)	Antenna type	Gain of the antenna (dBi)	Read range (cm)
3.85	Reader	2.6	74.72
3.94	Tag	3.8	83.83
5.8	Reader	1.7	44.72
5.65	Tag	1.25	43.59
8.12	Reader	2.7	35.84
8.2	Tag	2.65	35.28

Table 2. Comparison of results between the two methods MoM and FIT.

frequency (GHZ)	Calculation Method (Simulation software)	Gain of the antenna (dBi)	Read range (cm)
	READER ANTENNA		
3.6	FIT (CST)	3.4	87.50
3.85	MoM (FEKO)	2.6	74.72
5.8	FIT (CST)	2.7	50.10
5.8	MoM (FEKO)	1.7	44.72
8.2	FIT (CST)	2.6	35.06
8.10	MoM (FEKO)	2.7	35.84
	TAG ANTENNA		
3.9	FIT (CST)	3.9	85.60
3.94	MoM (FEKO)	3.8	83.83
5.9	FIT (CST)	3.0	51.02
5.65	MoM (FEKO)	1.25	43.59
8.2	FIT (CST)	2.2	33.48
8.2	MoM (FEKO)	2.65	35.28

6. COMPARISON WITH OTHER METHODS OF CALCULATION

This study has been done with The CST Microwave Studio which is based on the finite integral method (FIT) [3]. Our results are similar, but with small differences summarized in the Table 2.

Comparing these results, we note that:

- For the Reader, there is a small frequency lag of 0.25 GHZ ($f_{1\ FIT}$ = 3.6 GHZ and $f_{1\ MoM}$ = 3.85 GHZ)
- For the Tag, there is a small frequency lag of 0.25 GHZ ($f_{2\ FIT}$ = 5.9 GHZ and $f_{1\ MoM}$ = 5.65 GHZ)

7. CONCLUSIONS

The fractal antennas are preferred due to small size, light weight and multiband. This paper introduced RFID reader and tag antennas. The reader antenna exhibits a −10 dB bandwidth of 305 MHz at 3.85 GHz, 201 MHz at 5.8

GHz, and 289 MHz at 8.12 GHz. The tag antenna exhibits a −10 dB bandwidth of 165 MHz at 3.94 GHz, 149 MHz at 5.65 GHz, and 166 MHz at 8.2 GHz. The maximum read range obtained by the reader antenna is 74.72 cm, and 83.83 cm by the tag antenna.

By simulating the antennas using the Method of Moment (MoM) and comparing the results with another study using the method of finite integral (FIT), we find that the results are similar, even if there are small lags that are tolerable.

REFERENCES

1. H. Bhatt and B. Glover, "RFID Essentials," O'Reilly, Sebastopol, 2006.
2. S. B. Miles, S. E. Sharma and J. R. Williams, "RFID Technology & Applications," Cambridge University Press, New York, 2011.
3. C. Varadhan, J. K. Pakkathillam, M. Kanagasabai, R. Sivasamy, R. Natarajan and S. K. Palaniswamy, "Triband Antenna Structures for RFID Systems Deploying Fractal Geometry," IEEE Antennas and Wireless Propagation Letters, Vol. 12, 2013, pp. 437-440.
4. G. O. Young, "Synthetic Structure of Industrial Plastics (Book Style with Paper Title and Editor)," In: J. Peters, Ed., Plastics, McGraw-Hill, New York, 1964, pp. 15-64.

5. "FEKO 5.5 User's Manual," EM Software & Systems-S. A, 2009, pp. 1-1.

6. B. Bieda and P. Słobodzian, "Efficiency of the IE-MoM Approach in the Analysis of Dielectric Bodies Embedded in a Cavity," 2010 18th International Conference on Microwave Radar and Wireless Communications (MIKON), Vilnius, 14-16 June 2010, pp. 1-4.

7. A. Mehdipour, L. Rosca, A. Sabek, C. Trueman and S. Hoa, "Full Composite Fractal Antenna Using Carbon Nanotubes for Multiband Wireless Applications," IEEE Antennas Wireless Propagation Letters, Vol. 9, 2010, pp.891- 895.

8. A. Ismahayati, P. J. Soh and R. Hadibah, "Design and Ana- lysis of a Multiband Koch Fractal Monopole Antenna," 2011 IEEE International RF and Microwave Conference (RFM), Seremban, 12-14 December 2011, pp. 58-62.

9. D. H. Werner and S. Ganguly, "An Overview of Fractal Antenna Engineering Research," IEEE Antennas and Propagation Magazine, Vol. 45, No. 1, 2003, pp. 38-57.

10. A. Mak, C. R. Rowell, R. D. Murch and M. Chi-Lun, "Compact Multiband Planar Antenna for 2.4/3.5/5.2/5.8- GHz Wireless Applications," IEEE Antennas Wireless Propagation Letters, Vol. 11, 2012, pp. 144-147.

11. D. C. Chang, B. H. Zeng and J. Liu, "CPW-Fed Circular Fractal Slot Antenna Design for Dual-Band Applications," IEEE Transactions on Antennas and Propagation, Vol. 56, No. 12, 2008, pp. 3630-3637.

12. J. H. Gu, S. S. Zhong, L. L. Xue and Z. Sun "Dual-Band Monopole Antenna for 2.45/5.8 GHz RFID Applications," 2008 China-Japan Joint Microwave Conference, Shanghai, 10-12 September 2008, pp. 133-135.

13. B. D. Bala, M. K. A. Rahim, N. A. Murad, M. F. Ismail and H. A. Majid, "Design and Analysis of Metamaterial Antenna Using Triangular Resonator," 2012 Asia-Pacific Microwave Conference Proceedings (APMC), Kaohsiung, 4-7 December 2012, pp. 577-579.

3
CHAPTER

PERFORMANCE ANALYSIS OF OPTIMAL SCHEDULERS IN SINGLE CHANNEL DENSE RADIO FREQUENCY IDENTIFICATION ENVIRONMENTS

Javier Vales-Alonso*, Francisco Javier Parrado-García and Juan J Alcaraz

ABSTRACT

Schedulers in radio frequency identification dense environments aim at distributing optimally a set of t slots between a group of m readers. In single-channel environments, the readers within mutual interference range must transmit at different times; otherwise, interferences prevent identification of the tags. The goal is to maximize the expected number of tags successfully identified within the t slots. This problem may be formulated as a mixed integer non-linear mathematical program, which may effectively exploit available knowledge about the number of competing tags in the reading zone of each reader. In this paper, we present this optimization problem and analyze the impact of tag estimation in the performance achieved by the scheduler. The results demonstrate that optimal solutions outperform a reference scheduler based on dividing the available slots proportionally to the number of tags in each reader. In addition, depending on the scenario load, the results reveal that there exist an optimum number of readers for the topology considered, since the total average number of identifications depend non-linearly on the load. Finally, we study the effect of imperfect tag population knowledge on the performance achieved by the readers.

1 INTRODUCTION

Passive radio frequency identification (RFID) is increasingly being used to identify and trace objects in supply chains, in manufacturing process, and so forth. These environments are characterized by a large number of items with attached tags which flow on conveyor belts, inside pallets or boxes, and the like, entering and leaving facilities. In large realistic installations, several readers are commonly deployed; these are the so-called *dense reader environments*, comprising multiple readers within a mutual range.

In these scenarios, the rate of tags identified per reader is limited by the reader collision problems, namely:

- Reader-to-tag interferences (RTI) occur when two or more readers, irrespectively of the working frequency, transmit at the same time, overlapping their read ranges (reader-to-tag range) and powering the same tags. For instance, in Figure 1, if readers R and R' are feeding tag A simultaneously, tag is not able to produce a correct response to any of the readers.
- Reader-to-reader interferences (RRI) occur when two or more readers working at the same frequency are in mutual range, that is, one reader that powers a tag within its reader-to-tag range can receive stronger signals from other readers, ruining the weaker signal from the tag. For example, in Figure 1, tag B cannot be read by R if at the same time R' tries to read the tag C.

Another kind of interference is tag-to-tag interference which is internal to the reader's cell and is produced among tags competing to be identified by the reader. This latter type occurs even with a single reader, whereas the former ones (external) are only present with more than one reader. Indeed, the way of addressing internal and external interferences is completely different and independent. External ones are addressed by reserving (in real-time or with a preconfigured scheduling) resources to particular readers. Then, the reader uses these resources (time, frequency, power, etc.) to execute some algorithm to solve the tag-to-tag interference problem, as the static frame slotted ALOHA (static-FSA). Later in Section 2.1, we analyze the way how static-FSA and its derivate dynamic-FSA work, since their operation is relevant to decide the readers' scheduling.

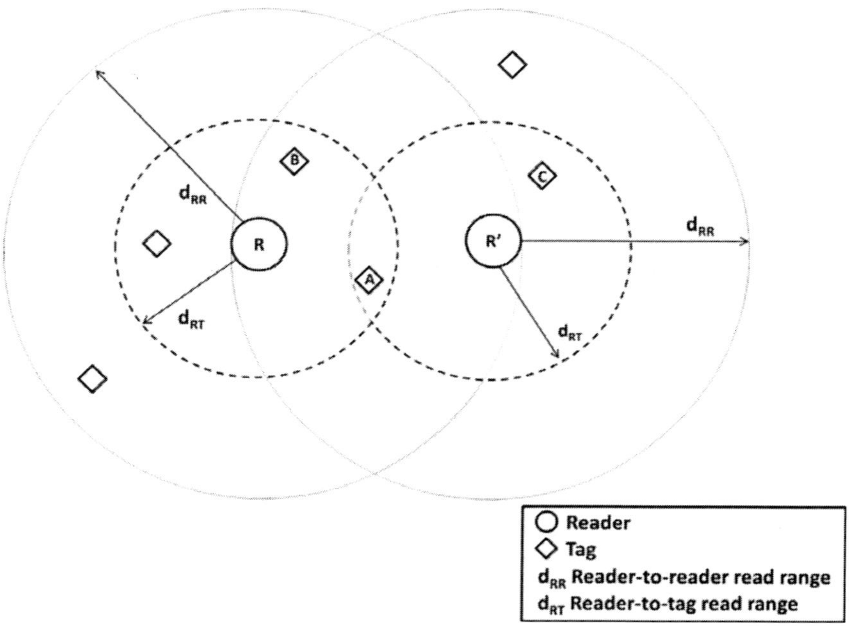

Figure 1 INTERFERENCES IN DENSE READER ENVIRONMENTS.

External interferences are directly related to the readers' output power, which delimit the interference range. For example, in Europe, output power may reach up to 2 W and guarantees a reader-to-tag range up to 10 m, while this may cause interferences with readers up to some hundreds of meters typically, determining interference ranges:

- If two or more readers are within two times the reader-to-tag range (d_{RT}), either part or the whole reading area overlaps, preventing tag operation. Hence, both RTI and RRI interferences are present. In this case, reader operation should be allocated at different working times.

- If the distance among the readers is between d_{RT} and the maximum distance determined by the RRI (d_{RR}), only RRI appears. The reader operation can be multiplexed either in frequency or in time.
- If the distance among the readers is larger than the maximum RRI distance, they do not suffer interferences.

Table 1 summarizes the restrictions applying to reader operation for dense reader environments.

Table 1 READER OPERATION RESTRICTIONS VERSUS **d**

	= Freq	≠ Freq
= Time	$d > d_{RR}$	$d > d_{RT}$
≠ Time	$d > 0$	$d > 0$

Therefore, in dense reader environments, the problem is how to distribute the reading resources available among the readers to perform optimally. The main parameters involved in this problem are the following:

- The number of readers, m.
- The number of available frequency channels, F.
- The number of time-slots available in each frequency, t.
- The topology of the readers.
- The implemented identification procedure in each reader (e.g., static-FSA, dynamic-FSA, Query-Tree protocols, etc.).
- The characteristics of the traffic of tags (e.g., static tags vs. tag flow, random vs. deterministic number of tags, etc.).

Current standards (see Section 2) propose some solutions to reduce collision issues but exclusively focused on minimizing RRI. On the other hand, a number of papers (see also Section 2) deal with minimization of the RTI but without considering reader-to-reader interferences.

In a previous paper [1], a particular simplified problem with two readers $m = 2$ in reader-to-reader range (*dual reader environment*) is addressed. Besides, in our previous paper [2], the scheduling problem for single-channel environments is firstly introduced, that is, we consider the case of any arbitrary number of readers (m) and for any particular network topology. Attending to the restrictions given above, in this case, the readers cannot transmit simultaneously if the reader-to-reader interferences are present, that is, if the distance between them is less than d_{RR} (note that this case also comprises reader-to-tag interferences).

In addition to [2], in this work, we provide insight on the impact of the schedulers derived from the knowledge of the tag population associated to each reader. To the best of our knowledge, all previous optimization models (see Section 2) have largely ignored the availability of this information. This information can be effectively exploited to construct a scheduler with the goal of maximizing the number of identifications in the whole interrogation period. In this work, we assume that this information is known and show how it can be used to develop an optimal scheduler. Moreover, we analyze the improvement obtained when this information is available and the effect on the expected performance when errors occur in tag estimation.

This resource allocation problem is addressed both for static and dynamic frame length identification procedures (which are described later in Section 2.1) and that the tags remain in coverage of their corresponding reader at least during the whole period of identification (t time-slots). The goal is to maximize the expected number of identified tags in the whole network.

The rest of the paper is organized as follows: In Section 2, the most relevant research proposals are shown. Section 2.1 describes the identification procedures commonly used in RFID readers. Section 3 describes the optimization model. Section 4 shows the performance results achieved by the optimal algorithm. Section 5 deals with the analysis of the impact of tag population estimation in the scheduler. Finally, Section 6 concludes and describes future works.

2 RELATED WORK

A number of proposal for coordinating dense reader environments have been presented in the literature; most of them are based on heuristic approaches and are, thus, suboptimal by nature. A summary of these works is contained in [3]. Besides, a number of papers [4–10] propose different system models and schedulers based on the optimization of some metric, defined upon the corresponding model.

Choi and Lee [4] propose a mixed integer linear program to minimize the reader interference problem as well as other performance metrics by selecting channel, timeslots, and output power for each reader. Their strategy is based on achieving a minimal signal-to-interference-plus-noise ratio for the signal received from tags, as well as on maximizing network utilization and minimizing power consumption. However, they neglect the availability of information about the number of tags present in the reading area of each reader and the operation of the underlying reading protocols, which are major factors determining the performance of the reading process.

Kim et al. [5] propose the TPC-CA algorithm based on a power control approach. It consists of controlling the reader output power optimally to reduce reader-to-reader collisions. Optimality criterion is related to minimize the area where interferences among readers occur.

Chui-Yu et al. introduces GA-BPSO in [6] a scheduler based on genetic algorithm and swarm intelligence meta-heuristics for single-channel environments. These schedulers aim at minimizing the overall sum of transaction times. However, these times are provided as parameters for the scheduler and are not based on the impact of resource allocation on the reading protocols.

Deolalikar et al. derive in [7] optimal scheduling schemes for readers in RFID networks for four basic configurations. As in our work, the authors aim at maximizing the number of identification within the scheduling period(t), but they model the performance of the reading process with an approximation: the number of tags identified increases linearly up to a saturation point. From that point on, the number of identifications remains constant. As we demonstrate in Section 2.1, this approach is not realistic for different tag-to-tag anti-collision protocol configurations (e.g., in static-FSA, there is a drop on the throughput). As in our work, only reader-to-reader interference (and thus single-channel) environments are considered.

The study of Mohsenian-Rad et al. [8] is the work more closely related to ours. The authors design two optimization-based distributed channel selection and randomized interrogation algorithms for dense RFID systems: FDFA (which is fully distributed and achieves a local optimum) and SDFA (semi-distributed and reach to the global optimum). In addition, the authors realistically assume that the reader may operate asynchronously. Similarly to our work, they consider a FDMA/TDMA scheduler, where the medium access control layer of the readers complies with EPCglobal Class-1 Gen-2 standard (therefore, a reader may allocate a number of interrogation frames within its allocated time). In this work, the authors focus on the probability that a reader starts an interrogation interval without experiencing either reader-to-reader or reader-to-tag collisions. The goal is to achieve max-min fairness in the network; as a result, the processing load is evenly distributed among all readers. However, this paper does not consider the knowledge about the number of contending tags in range of each reader. This information allows us to formulate the optimization problem in terms of reading efficiency (maximizing the number of tags identified in the overall time period). An additional contribution of [8] is to develop a protocol to construct the topology (i.e., reader-to-reader and reader-to-tag constraints) by exchanging some messages in three control channels. This protocol may be implemented in other schedulers (like ours) to determine the network topology in real time.

Tanaka and Sasase [9] also determine an interference model which they apply later to formulate constraints in a binary integer linear program aimed at maximizing the ratio of total time where readers can successfully communicate with the tags and total interrogation time of the readers. As in our model, the goal is selecting suitable timeslot and channels for each reader. They also propose two heuristics (one distributed and one centralized) to solve the allocation problem efficiently.

Seo and Lee [10] propose a FDM/TDM scheduler (RA-GA) based on a reader-to-reader interference model, which seeks to maximize a utility function depending upon the operating time slots. This problem is solved using a genetic algorithm meta-heuristic.

As many of the previous works, neither in [9] nor in [10] the reading protocol or the current load (unidentified tags) of each cell is considered. Summarizing, to the best of our knowledge, all previous optimization models ignore the availability of information about the number of tags within the range of each reader. This information can be very effectively exploited to construct a scheduler with the goal of maximizing the number of identifications in the interrogation period. Besides, most previous works assume a model view from the physical layer perspective and are usually aimed at minimizing interference. This view has notable limitations since tag identification performance, and thus scheduling, heavily depends on the underlying tag-to-tag anti-collision protocol, as discussed in the next Section.

2.1 Tag identification procedure

The identification process involves communications between the reader and the tags and takes place in a shared wireless channel. Basically, the reader *interrogates* tags nearby by sending a *Query* packet (the exact format of this packet depends on the particular standard). The tags are energized by the reader's signal and respond to this request with their identification.

When several tags answer simultaneously, a tag-to-tag collision occurs, and the information cannot be retrieved. Therefore, a tag-to-tag anti-collision mechanism is required when multiple tags are in range. ALOHA-based protocols, also called probabilistic or random access protocols, are the most prevalent in the UHF band. They are designed for situations in which the reader does not know exactly how many tags will cross its checking area. The most common ALOHA RFID protocol is FSA, a variation of slotted ALOHA. As in slotted ALOHA, time is divided into time units called *slots*. However, in FSA, the slots are subject to a super-structure called a 'frame'. Two options of the FSA are commonly used in the RFID technology:

1. *static frame length FSA (static-FSA)*. The reader starts the identification process with an identification frame by sending a Query packet with information about the frame length (k slots) to the tags. The frame length is kept unchanged during the whole identification process. At each frame, each unidentified tag selects a slot at random from among the k slots to send its identifier to the reader. FSA achieves reasonably good performance at the cost of requiring a central node (the reader) to manage slot and frame synchronization. FSA has been implemented in many commercial products and has been standardized in the ISO/IEC 18000-6C [11], ISO/IEC 18000-7 [12], and EPCGlobal Class-1 Gen-2 (EPC-C1G2) standards [13].

2. *dynamic frame length FSA (dynamic-FSA)*. When the tags outnumber the available slots, the identification time increases considerably due to frequent tag-to-tag collisions. On the other hand, if the slots outnumber the tags, many slots will be empty in the frame, which also leads to long identification times. Dynamic-FSA protocols were conceived to address this problem. They are similar to static-FSA, but the number of slots per frame is variable. In other words, parameter k may change from frame to frame in the Query packet to adjust the frame length. Dynamic-FSA operation is optimal in terms of reading throughput (rate of identified tags per slot) when the frame length equals the number of contenders [14]. Therefore, to maximize throughput, the reader should ideally know the actual number of competing tags and allocate that number of slots to the next frame. Different dynamic-FSA algorithms have been proposed to estimate the number of competing nodes based on the collected statistical information. The most relevant ones have been studied in depth in our previous papers [15, 16].

In the next Section, both algorithms (static-FSA and dynamic-FSA) are considered in order to propose an optimal slot distribution for the single channel environment. In the case of static-FSA, the frame length is k for all readers; in the case of dynamic-FSA, we are assuming that each reader j actually knows the number of competing nodes at frame i ($n_{j,i}$) and that the reader is adjusting $k_{j,i} = n_{j,i}$ if the number of the remaining available slots is greater than $n_{j,i}$.

3 OPTIMAL TIME DISTRIBUTION

Recall from the introduction that a dense-reader environment with the limitation of a single frequency channel $F = 1$, m readers, and t slots available in the channel is assumed. In addition, for each reader $j = 1, \ldots, m$, let us denote

- n_j, the tags unidentified in the range of the reader j
- t_j, the number of slots assigned to reader j.

Let us remark that the methods used in dynamic-FSA tag-to-tag anti-collision protocols to determine the number of contenders can be directly applied in our case to estimate n_j in real-time (see [15] and [16] for details). Besides, topological dependencies among readers are defined by an $m \times m$ matrix $A=(ajj')$, the elements of which are 1 if reader j and j' cannot operate at the same time, and 0 otherwise.

Let $\varphi(n,t)$ denote the expected number of identified tags when n tags contend in t slots, and let us define Φ as the whole expected number of identified tags in the network, that is,

$$\Phi = \sum_{j=1}^{m} \varphi(n_j, t_j). \tag{1}$$

Then, the optimization problem can be stated as solving

$$\max_{\substack{t_j \\ j=1,\ldots,m}} \Phi. \tag{2}$$

Subject to

$$t_j \geq 0 \tag{3}$$

and

$$t_j + I_{t_j} \sum_{j'=1, j' \neq j}^{m} t_{j'} a_{j'j'} \leq t \ \text{for all } j = 1, \ldots, m, \tag{4}$$

where Itj is 1 if t_j is greater than 0, and 0 otherwise.

Constraint (3) expresses a basic limiting condition on the values assigned to the number of assigned slots. The key in our problem formulation is constraint (4) which establishes local conditions to regulate the spatial reuse of the resources in our network. This condition states that the number of slots assigned to a reader j plus those assigned to its neighbors can not surpass the number of available slots. Itj

is included since readers without slots assigned should be considered as disconnected, and no constraints have to be applied to that particular readers.

The former constraint guarantees that enough slots are available for each node in each neighborhood (set of nodes bonded with topological constraints, i.e., aj′j=1

) to obey with the limit of t slots among all neighbors. Note that it does not guarantee that these slots can be allocated consecutively. However, this is not an issue since tags do not proceed with the next slot until a *QueryRep* packet arrives from the reader. Hence, even if the slots are not consecutively allocated, the tags perceive continuity and the identification can be performed seamlessly.

We must remark that this set of constraints produces feasible solutions regardless of the considered topology. However, in some cases (as we will show in the next section), the constraint is too strict and may lead to suboptimal solutions since space reutilization is limited. If the network graph has a large density (i.e., the number of edges is close to the maximal number of edges), the results provided by solving problem (1) will be close to the optimal solution with maximal space resource reutilization. Whereas, for sparse network graphs, the space reutilization will be small. The first kind of scenario will likely occur (due to the large reader-to-reader interference range) in facilities with non-screened readers; thus, the solutions found will be realistic.

3.1 $\varphi(n_j, t_j)$ computation for static-FSA

Finally, in order to solve the optimization problem, the expected number of identifications $\varphi(n,t)$ must be computed. The next sections deal with its computation both for static-FSA and for dynamic-FSA.

In this case, the reading process for each reader j consists of several consecutive reading frames of length k until all the t_j reading slots are eventually exhausted. It is assumed that $t_j = k\ a$, being a a positive integer or zero. Given the last condition, and since expectation is a linear operator, $\varphi(n_j,t_j)$ can be computed as the sum of the average number of tags identified in the first frame ($\varphi(n_j,k)$) plus those identified in the remainder process ($\varphi(n_j-\eta,t_j-k)$), where η denotes the random number of tags identified in the first frame.

The former part can be computed if the distribution of the random variable η is known; so, let us denote $P(a|n_j,k)$ as the probability that a tags are identified if n_j tags contend in k slots. Then

$$\varphi(n_j, k) = \sum_{a=0}^{n_j} aP(a|n_j, k). \tag{5}$$

The second part, can be computed given the joint probability of identifying a tags in the first frame and a' in the remainder process if n_j tags contend in t_j slots, which we will denote as $P(a,a'|n_j,t_j)$. We obtain

$$\varphi(n_j - \eta, t_j - k) = \sum_{a=0}^{n_j} \sum_{a'=0}^{n_j} a' P(a, a'|n_j, t_j). \tag{6}$$

But, clearly $P(a, a'|n_j, t_j) = P(a|n_j, k)P(a'|n_j - a, t_j - k)$, and this leads to

$$\varphi(n_j - \eta, t_j - k) = \sum_{a=0}^{n_j} \varphi(n_j - a, t_j - k)P(a|n_j, k) \tag{7}$$

Appendix 1 demonstrates that the value of $P(a|n_j,t_j)$ is given by (where the technique in [17] was used to compute the probability $P(a|n,t)$)

$$P(a|n_j, t_j) = \frac{n_j!}{t_j^{n_j}} \binom{t_j}{a} \sum_{c=0}^{n_j - a} (-1)^c \binom{t - a}{c} \frac{(t_j - a - c)^{n_j - a - c}}{(n_j - a - c)!}. \tag{8}$$

Thus, the following recursive equation results:

$$\varphi(n_j, t_j) = \begin{cases} \sum_{a=0}^{n_j} (a + \varphi(n_j - a, t_j - k)) P(a|n_j, k), \text{ if } t_j \geq k \\ 0, \text{otherwise} \end{cases} \tag{9}$$

3.2 $\varphi(n_j, t_j)$ computation for dynamic-FSA

In this second case, the reading process for each reader j also consists of several reading frames but of variable length $k_{j,1}, k_{j,2}, \ldots$, until all the t_j reading slots are exhausted. Besides, denote the number of contenders in each frame as $n_{j,1}, n_{j,2}, \ldots$. Since the dynamic-FSA operation is used (see Section 2.1), the reader seeks to maximize reading throughput and allocates the optimal number of slots in each frame. That is, as much slots as the number of contending tags ($k_{j,i} = n_{j,i}$). This is possible while $n_{j,i} < t_j - \sum_{c=1}^{i-1} k_{j,c}$, that is, if the remainder number of slots is greater that the number of contenders. Otherwise, we assume that a last frame is allocated with all the remaining slots $(k_{j,i} = t_j - \sum_{c=1}^{i-1} k_{j,i})$.

Like in the previous case $\varphi(n_j, t_j)$ can be described through a recursive equation:

$$\varphi(n_j, t_j) = \begin{cases} \varphi(n_j, n_j) + \sum_{a=0}^{n_j} \varphi(n_j - a, t_j - n_j) P(a|n_j, n_j), \text{ if } n_j < t_j \\ \varphi(n_j, t_j) \text{ if } n_j \geq t_j. \end{cases}$$

From Equation (5),

$$\varphi(n_j, n_j) = \sum_{a=0}^{n_j} a P(a|n_j, n_j),$$

and

$$\varphi(n_j, t_j) = \sum_{a=0}^{n_j} aP(a|n_j, t_j), \text{if } n_j \geq t_j.$$

Hence,

$$\varphi(n_j, t_j) = \begin{cases} \sum_{a=0}^{n_j} (a + \varphi(n_j - a, t_j - n_j)) P(a|n_j, n_j) \text{ if } n_j < t_j \\ \sum_{a=0}^{n_j} aP(a|n_j, t_j) \text{ if } n_j \geq t_j. \end{cases} \quad (10)$$

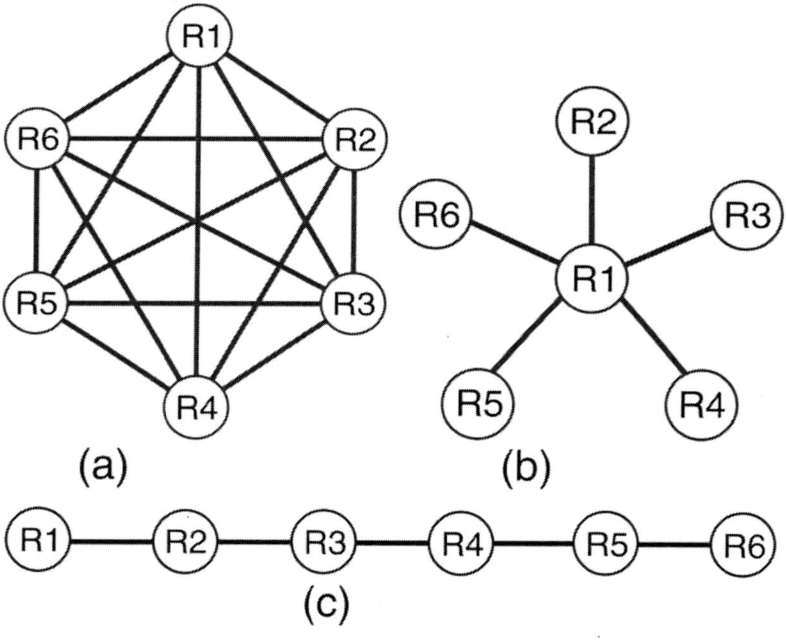

Figure 2 EXAMPLE SCENARIOS FOR m **= 6.** (**A**) Full-mesh topology. (**B**) Star topology. (**C**) Line topology.

4 RESULTS

The optimal assignment has been computed in static-FSA and dynamic-FSA cases using the recursive formulas described in the previous section. Three representative scenarios (see Figure 2) have been selected. The edges (i.e., connecting lines) represent the existence of interference between two vertices (readers). On the first scenario, a full-mesh topology of *m* readers has been selected. It is a typical configuration in facilities, since the RRI distance is large (in the order of hundreds of meters) as discussed in the introduction. On the other hand, the star topology of m readers selected for scenario two represents another practical case, where readers are confined to some areas (e.g., by screening the reading area), and interferences are restricted to some particular pairs, exclusively between R1 and the other readers in this example. Finally, the line scenario is representative of an assembly line, where neighbor readers are in range.

Besides, the following parameters have been considered:

- $t = 512$,
- n tags to be identified at each reader, from 1 to 100 tags,
- $m = 2, 4, 6, 8$, and 10,
- and for static-FSA $k = 16$ and 64.

Our optimization algorithm has been implemented using the *General Algebraic Modeling System*, a high-level modeling system for mathematical programming and optimization, and AlphaECP, a MINLP (Mixed-Integer Non-Linear Programming) solver based on the extended cutting plane method. It allowed us to define our optimization problem directly from the mathematical description provided in Section 3.

Tables 2, 3, and 4 show the optimal configurations (slots assigned to each reader) for the dynamic-FSA protocol in all scenarios with $m = 4$. Let us remark that the optimal solutions are non-trivial, that is, can be obtained through an educated guess. Clearly, this solution is not unique: a circular permutation of the optimal solution, replacing the slots from Rj to R $(j + 1)$ if $j < m$, and from Rm to R1 is also an optimal solution for the full-mesh scenario. The same applies to the star scenario if the slots in R1 are kept constant while any permutation is applied to the rest of the readers and to the line scenario replacing the slots from R_j to R $(m\text{-}j)$.

In addition, note that the results obtained for the star scenario (Table 3) can be improved. For example, for $n = 30$, after assigning 128 slots to R1, it would be possible to assign 384 to all remainder readers, which will provide a solution better than that obtained by solving problem (1). As discussed in Section 3, this is caused by the strict resource reutilization obtained by applying the set of constraints given by Equation (4). This problem does not appear for networks characterized by a dense graph, as the full-mesh scenario.

Table 2 FULL-MESH SCENARIO

Number of tags	Φ	R1	R2	R3	R4
10	40.000	128	128	128	128
20	80.000	126	115	155	116
30	119.999	128	128	128	128
40	159.427	128	128	128	128
50	186.355	128	128	128	128
60	189.195	94	140	139	139
70	189.113	110	70	166	166
80	188.949	80	162	190	80
90	188.992	0	212	89	211
100	188.905	157	157	99	99

Optimal assignment of slots for the dynamic-FSA protocol.

Table 3 STAR SCENARIO

Number of tags	Φ	R1	R2	R3	R4
10	40.000	78	74	76	77
20	80.000	126	115	155	116
30	119.999	128	128	128	128
40	159.427	128	128	128	128
50	186.355	128	128	128	128
60	189.195	94	140	139	139
70	210.000	0	512	512	512
80	240.000	0	512	512	512
90	270.000	0	512	512	512
100	300.000	0	512	512	512

Optimal assignment of slots for the dynamic-FSA protocol.

Table 4 LINE SCENARIO

Number of tags	Φ	R1	R2	R3	R4
10	40.000	85	86	165	221
20	80.000	118	118	124	248
30	120.000	153	153	171	171
40	160.000	173	170	169	173
50	199.913	174	169	169	174
60	236.153	176	168	168	176
70	254.481	188	162	162	188
80	263.693	215	172	125	215
90	273.451	242	180	90	242
100	287.462	256	256	0	453

Optimal assignment of slots for the dynamic-FSA protocol.

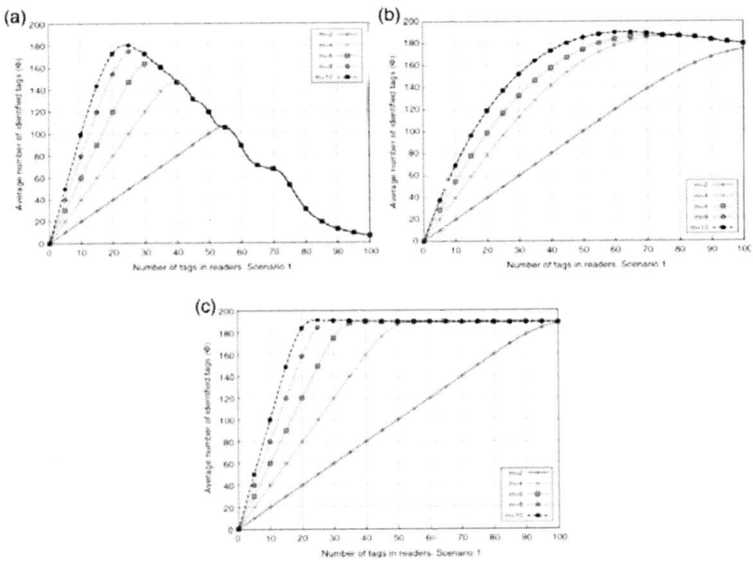

Figure 3 EXPECTED NUMBER OF IDENTIFICATIONS (Φ) VERSUS **n** : FULL-MESH SCENARIO. (**A**) Static-FSA $k = 16$. (**B**) Static-FSA $k = 64$. (**C**) Dynamic-FSA.

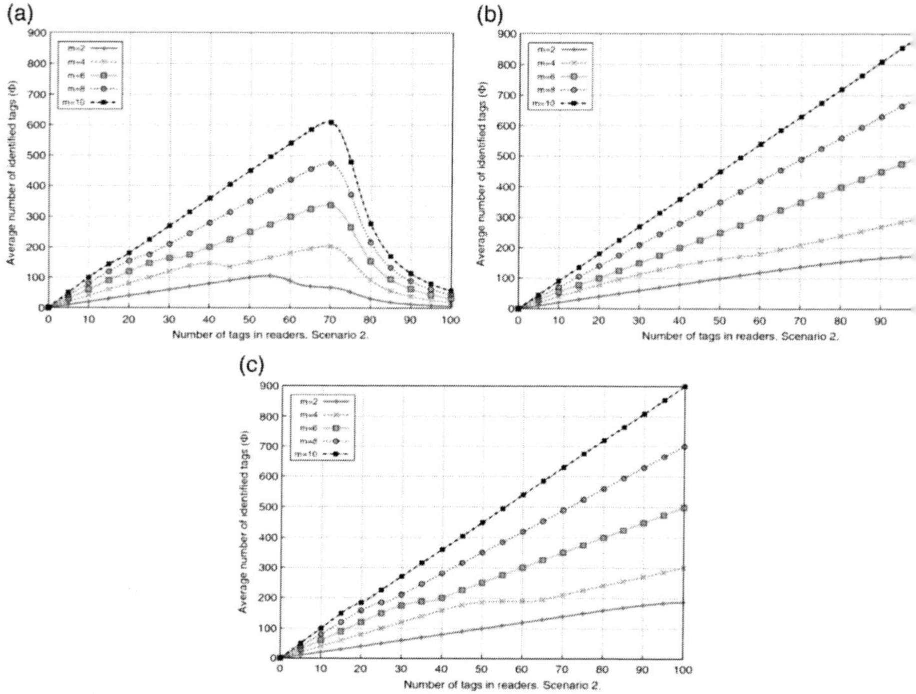

Figure 4 EXPECTED NUMBER OF IDENTIFICATIONS (Φ) VERSUS **n** : STAR SCENARIO. **(A) Static-FSA *k* = 16. (B) Static-FSA *k* = 64. (C) Dynamic-FSA.**

Besides, Figures 3, 4, and 5 show the expected number of tags identified (Φ) for all the possible values of *m* using the optimal assignments. Note that the resources available ($t = 512$) are the same for all the configurations; however, the performance clearly varies. This illustrates how the underlying reading protocol determines the final system performance. Dynamic-FSA performs better than static-FSA assignment for both configurations of *k* (16, 64), as can be expected. This is reasonable since dynamic-FSA achieves an optimal reading throughput frame-by-frame while the number of available slots is at least equal to the number of contenders.

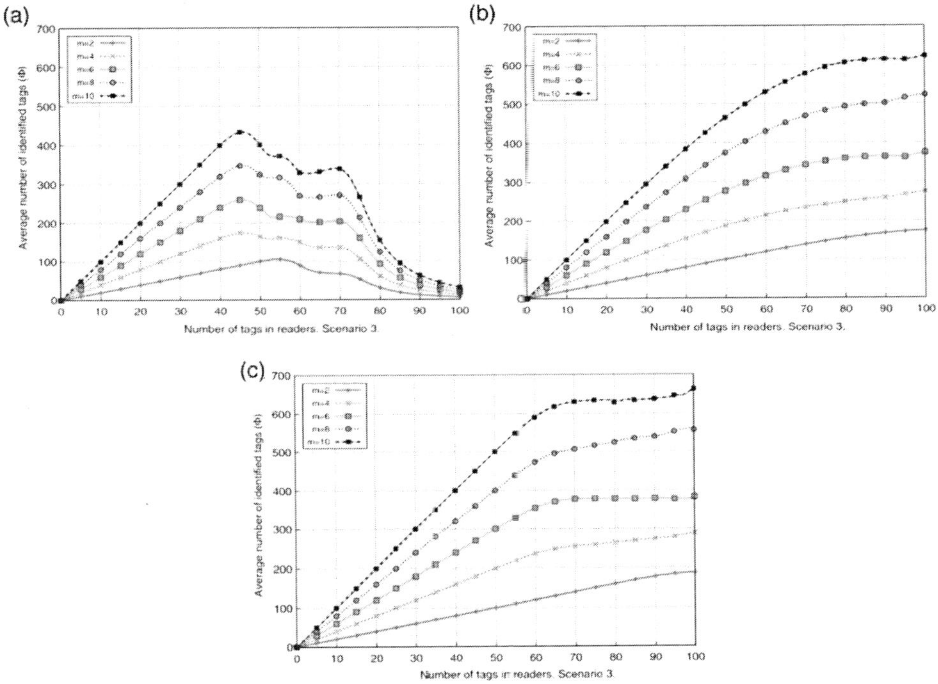

Figure 5 EXPECTED NUMBER OF IDENTIFICATIONS (Φ) VERSUS **n** : LINE SCENARIO. **(A) Static-FSA k = 16. (B) Static-FSA k = 64. (C) Dynamic-FSA.**

Another important result shown in these figures is the existence of saturation points in the system. That is, in some cases, the throughput does not increase when the load is increased. For dynamic-FSA, in all cases, the throughput never decreases; this is caused by the flexibility of dynamic-FSA to adapt to different loads. For static-FSA k = 64, the effect is almost similar to the dynamic-FSA case, except in the full-mesh scenario where the throughput slightly decreases when the tag is beyond $n = 60$. However, for all static-FSA k = 16 cases, the effect of the load in the throughput is dramatic, with a throughput minima and a step decreasing performance. This is of considerable importance, since static-FSA $k = 16$ is the default configuration of many readers in the market, and this configuration leads to poor collective performance.

In addition Figures 6, 7, and 8 show, for $m = 4$, the performance of the optimal allocation versus a non-optimal allocation scheme selected for comparison, namely, using 1m

of time allocated to each reader ('proportional' resource sharing), that is, $t_1 = ... = t_4 = 128$. This heuristic is a natural choice, since the number of tags in the range of each reader is the same; therefore a good performance could be expected. In fact, the proportional scheme achieves in a range of n a performance nearly equal to the optimal one, as can intuitively be expected, in some cases (e.g., static-FSA $k = 64$ and dynamic-FSA in the full-mesh topology). However, for some cases, the allocation is clearly suboptimal (e.g., star and line scenarios for $n > 60$). Noteworthy, in the star scenario, there is a point ($n \geq 70$) where the best option is directly to disconnect the central reader. In this case, without restrictions in the network, the remainder readers can be allocated each all the 512 slots. A similar behavior occurs for the line scenario, disconnecting some readers when there is a high load (e.g., see Table 4; if $n = 100$ the third reader is disconnected).

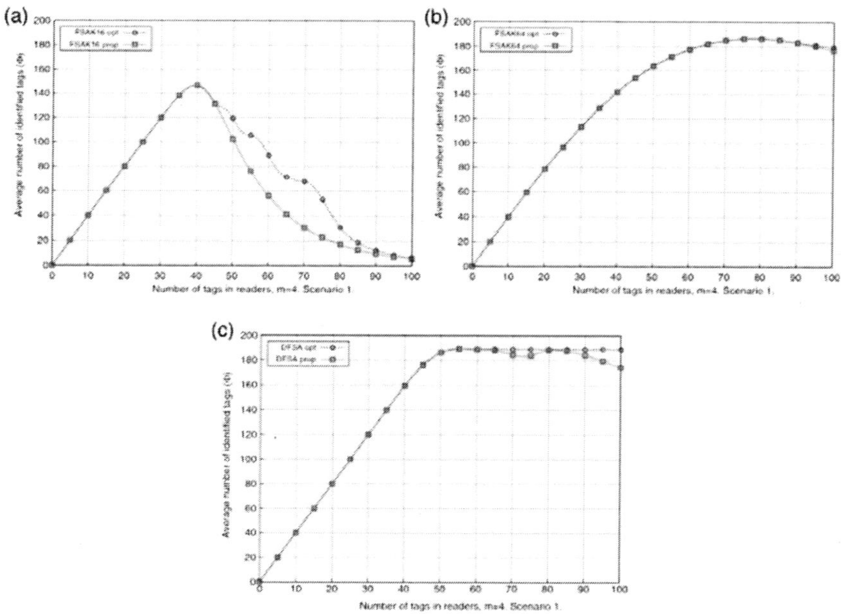

Figure 6 OPTIMAL VS. PROPORTIONAL ALLOCATION FOR FULL-MESH TOPOLOGY, **m** $= 4$. (A) **Static-FSA** $k = 16$**.** (B) **Static-FSA** $k = 64$**.** (C) **Dynamic-FSA.**

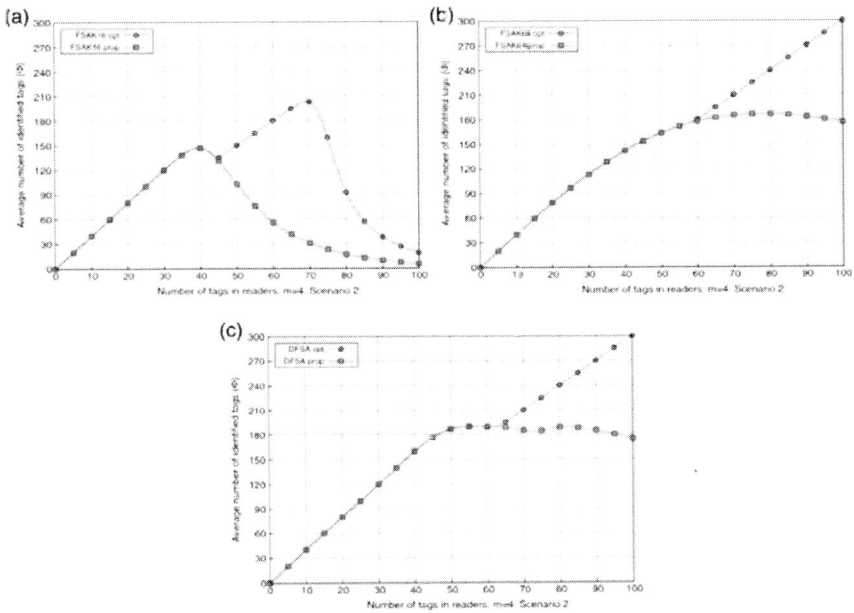

Figure 7 OPTIMAL VS. PROPORTIONAL ALLOCATION FOR STAR TOPOLOGY, **m** = 4. (A) **Static-FSA** $k = 16$**.** (B) **Static-FSA** $k = 64$**.** (C) **Dynamic-FSA.**

5 TAG ESTIMATION IMPACT ON SCHEDULER PERFORMANCE

The aim of this section is twofold:

1. Quantify the improvement achieved in the scheduler when tag instant population estimation is available.
2. Quantify the impact of tag population estimation errors on the performance achieved by the scheduler.

As stated in Section 2, previous works do not assume knowledge about the tag population and are mostly based on minimizing interferences. To establish a comparison between our model and a reference model that do not use population information, at least, we must focus on the same performance metric, i.e., the expected number of identifications (which can also be viewed as throughput).

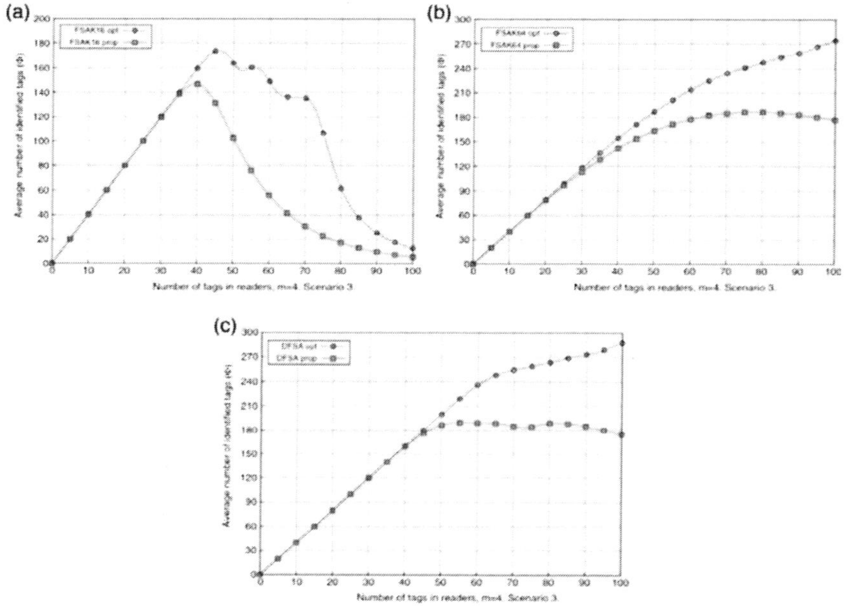

Figure 8 OPTIMAL VS. PROPORTIONAL ALLOCATION FOR LINE TOPOLOGY, **m** = 4. **(A) Static-FSA *k*= 16. (B) Static-FSA *k*= 64. (C) Dynamic-FSA.**

Although our reference model does not use information about the instant population, it is rational to assume at least a coarse knowledge of the environment, typically the average number of competing tags. This allows the designer to configure the system for a standard case. Note that if this information is unavailable, the designer should guess somehow a configuration, and the performance would be lower than in the reference model.

Henceforth, let us assume that our reference model is based on the availability of information about the average tag population and that the designer is able to select the optimal scheduler configuration for this case (e.g., by solving problem (1)).

For simplicity, let us denote by $\vec{n}$

the m-dimensional vector $(n_1,\ldots,n_m)$, and by $\Phi\vec{n}(n'\rightarrow)$ the expected number of identifications when the optimal solution to problem (1) with tag estimation parameter $\vec{n}$ is applied to the actual population $n'\rightarrow$. Besides, let $n*-\rightarrow$

denote the m-dimensional vector, where the j th component is the average number of tags in reader j.

Thus, if the probability distribution of tag population, i.e., $P(\vec{n})$

is known, the improvement (Δ) achieved by our scheduler over the reference scheduler is

$$\Delta = \sum_{\forall \vec{n}} (\Phi_{\vec{n}}(\vec{n}) - \Phi_{\vec{n}^*}(\vec{n}))P(\vec{n}). \tag{11}$$

By solving problem (1), both optimal slot assignments can be computed. Let us denote $\hat{t}$ and $\hat{t^*}$ to the optimal assignments for tag populations $\vec{n}$ and $\vec{n^*}$, respectively, and $\hat{t}_j$ and $\hat{t^*}_j$ the slots assigned to particular reader j th.

Then, Δ can be rewritten as

$$\Delta = \sum_{\forall \vec{n}} \left(\sum_{j=1}^{m} \varphi(n_j, \hat{t}_j) - \sum_{j=1}^{m} \varphi(n_j, \widehat{t^*}_j) \right) P(\vec{n}), \tag{12}$$

where $\varphi(n,t)$ is computed directly with formulas (9) and (10) for static-FSA and dynamic-FSA, respectively.

In addition, it could be argued that instant tag population estimation may be subject to errors. This can be included in our computations through an error vector $\vec{\epsilon}$, where the j th component is $\epsilon_j = \check{n}_j - n_j$, n_j the estimation, and $\check{n}_j$ is the actual number of tags. Therefore, the real tag distribution is $\vec{n} + \vec{\epsilon}$, and Δ should be modified as

$$\Delta = E_{\vec{\epsilon}} \left\{ \sum_{\forall \vec{n}} \left(\sum_{j=1}^{m} \varphi(n_j y + \epsilon_j, \hat{t}_j) - \sum_{j=1}^{m} \varphi(n_j + \epsilon_j, \widehat{t^*}_j) \right) P(\vec{n}) \right\}. \tag{13}$$

Note that in this last case, $\hat{t}_j$, is still computed with the estimation vector $\vec{n}$. So, if $\vec{n}$ and $\vec{\epsilon}$ are independent, we finally reach to

$$\Delta = \sum_{\forall \vec{\epsilon}} \left[\sum_{\forall \vec{n}} \left(\sum_{j=1}^{m} \varphi(n_j + \epsilon_j, \hat{t}_j) - \sum_{j=1}^{n} \varphi(n_j + \epsilon_j, \widehat{t^*}_j) \right) P(\vec{n}) \right] P(\vec{\epsilon}) \tag{14}$$

Note that we use a perfect knowledge of the average number of tags; therefore, we are assuming the least-favorable comparison case for our scheduler versus the reference model.

5.1 Numerical examples

For the sake of example, let us assume that for each reader, the number of tags is given by a uniformly distributed random variable **N** in the range [0,100]. That is, $n_j = \mathbf{N}$ for all $j = 1,\ldots,m$. Hence, $\overrightarrow{n*}=(50,\ldots,50)$. Tables 5, 6, and 7 show the average performance improvement achieved for the examples described in the previous section and for the dynamic-FSA and static-FSA tag-to-tag anti-collision protocols. The results are shown as the ratio of improvement (Δ) to the expected readings without using tag estimation.

Table 5 FULL-MESH SCENARIO

Number of readers	Dynamic-FSA	Static-FSA, $k = 64$	Static-FSA, $k = 16$
2	0.0273	0	0.0889
4	0.0120	0.0008	0.2891
6	0.0423	0.0050	0.4831
8	0.0791	0.0086	0.6024
10	0.0947	0.0086	0.6731

Ratio of improvement using tag estimation. $\epsilon = 0$.

Table 6 STAR SCENARIO

Number of readers	Dynamic-FSA	Static-FSA, $k = 64$	Static-FSA, $k = 16$
2	0.0273	0	0.0889
4	0.1592	0.1702	0.2636
6	0.0509	0.0023	0.1940
8	0.0417	0.0005	0.1756
10	0.0333	0.0000	0.1683

Ratio of improvement using tag estimation. $\epsilon = 0$.

Table 7 LINE SCENARIO

Number of readers	Dynamic-FSA	Static-FSA, $k = 64$	Static-FSA, $k = 16$
2	0.0266	0	0.0817
4	0.0331	0.0144	0.2114
6	0.0049	0.00[illegible]8	0.2105
8	0.0298	0.0050	0.2073
10	0.0101	0.0013	0.2056

Ratio of improvement using tag estimation. $\epsilon = 0$.

The results clearly depend on the scenario and on the tag-to-tag anti-collision protocols. Improvement ranges from nearly 0% in many static-FSA k=64 cases, while it may reach up to 67% for static-FSA, k=16 in the full-mesh scenario. For dynamic-FSA, the improvement is between 2.7% and 21.14%, depending on the particular scenario. Let us remark again that this comparison is performed against the average tags identified when the optimal configuration is computed using as information the mean number of competing tags. Therefore, this is the *minimum* improvement ratio: non-optimal schedulers (as the reference heuristic used in Section 4) will obtain worse results.

Besides, we can consider the estimation error. In our test we have assumed for each reader an error distributed uniformly $\epsilon \sim U[-10,10]$. Results are shown in Tables 8, 9, and 10.

Table 8 FULL-MESH SCENARIO

Number of readers	Dynamic-FSA	Static-FSA, $k = 64$	Static-FSA, $k = 16$
2	0.0284	0	0.0527
4	0.0064	0.0007	0.2334
6	0.0361	0.0040	0.4107
8	0.0580	0.0101	0.5515
10	0.0747	0.0070	0.5712

Ratio of improvement using tag estimation. $\epsilon \sim U[-10, 10]$.

Table 9 STAR SCENARIO

Number of readers	Dynamic-FSA	Static-FSA, $k = 64$	Static-FSA, $k = 16$
2	0.0275	0	0.0704
4	0.1541	0.1750	0.2087
6	0.0543	0.0017	0.2123
8	0.0406	-0.0002	0.1878
10	0.0302	0.0000	0.1925

Ratio of improvement using tag estimation. $\epsilon \sim U[-10, 10]$.

Table 10 LINE SCENARIO

Number of readers	Dynamic-FSA	Static-FSA, $k = 64$	Static-FSA, $k = 16$
2	0.0442	-0.0006	0.0023
4	0.0324	0.0212	0.0991
6	0.0003	0.0019	0.1169
8	0.0279	0.0071	0.0815
10	0.0118	0.0012	0.0942

Ratio of improvement using tag estimation. $\epsilon \sim U[-10, 10]$.

Again, the results heavily depend on the configuration, but in most cases, even assuming an error in the tag number estimation, they show a positive feedback using the estimation. In some cases, in the full-mesh and line scenarios, there is a negative impact, but almost negligible. Therefore, we can conclude that even assuming errors, the utilization of tag estimators is worth to be considered.

6 CONCLUSIONS

This work introduced a novel optimal scheduler for a particular dense reader environment composed by *m* readers which must share a single frequency channel. The scheduler proposed exceeds in performance to heuristic algorithms, improving the average number of tags identified in an RFID facility. Besides, the effect of the reading protocols has also been studied in depth, concluding that a dynamic FSA algorithm excels static frame length ones. Indeed, the impact of using knowledge about tag population in the scheduler has been analyzed. It has been concluded that even assuming errors in the estimation, our scheduler is able to obtain a higher performance than a reference model, where the average population is perfectly known.

As future works, we aim at extending our model to multi-channel scenarios, developing a model that allow full resource reutilization and further analyze RFID realistic scenarios to propose optimal configuration strategies.

REFERENCES

1. Vales-Alonso J, Bueno-Delgado MV, Alcaraz JJ: ***OPTIMAL SCHEDULING IN DUAL READER RFID ENVIRONMENTS.*** In Paper presented in 2011 IEEE international conference on RFID-technologies and applications (RFID-TA). Spain: Sitges; 15–16 September 2011.
2. Vales-Alonso J, Parrado-Garcia FJ, Alcaraz JJ, Egea-Lopez E: ***OPTIMAL SCHEDULING IN SINGLE CHANNEL DENSE READER RFID ENVIRONMENTS.*** In Paper presented in 2012 fourth international EURASIP workshop on RFID technology (EURASIP RFID). Torino, Italy; 27–28 September 2012.
3. Bueno-Delgado MV, Vales-Alonso J, Angerer C, Rupp M: ***A COMPARATIVE STUDY OF RFID SCHEDULERS IN DENSE READER ENVIRONMENTS.*** In Paper presented in IEEE international conference on industrial technology. Viña del Mar, Chile; 14–17 March 2010.
4. Choi J, Lee C: ***AN MILP-BASED CROSS-LAYER OPTIMIZATION FOR A MULTI-READER ARBITRATION IN THE UHF RFID SYSTEM. SENSORS.*** 2011,***11***(3):2347-2368.

5. Kim J, Lee W, Kim E, Kim D, Suh K: ***OPTIMIZED TRANSMISSION POWER CONTROL OF INTERROGATORS FOR COLLISION ARBITRATION IN UHF RFID SYSTEMS. IEEE COMMUN. LETT.*** 2007,***11***(1):22-24.

6. Chui-Yu C, Cheng-Hsin K, Chen KY: ***OPTIMAL RFID NETWORKS SCHEDULING USING GENETIC ALGORITHM AND SWARM INTELLIGENCE.*** In Paper presented in IEEE international conference on systems, man and cybernetics,. San Antonio, USA; 11–14 October 2009.

7. Deolalikar V, Recker J, Mesarina M, Pradhan S: Optimal scheduling for networks of RFID readers Paper presented at the first international workshop on RFID and ubiquitous sensor networks. Nagasaki,Japan: EUC Workshops LNCS 3823); 6–9 December 2005.

8. Mohsenian-Rad AH, Shah-Mansouri V, Wong VWS, Schober R: ***DISTRIBUTED CHANNEL SELECTION AND RANDOMIZED INTERROGATION ALGORITHMS FOR LARGE-SCALE AND DENSE RFID SYSTEMS. IEEE TRANS. ON WIRELESS COMMUN.*** 2010,***9***(4):1402-1413.

9. Tanaka Y, Sasase I: ***INTERFERENCE AVOIDANCE ALGORITHMS FOR PASSIVE RFID SYSTEMS USING CONTENTION-BASED TRANSMIT ABORTION. IEICE TRANS. COMMUN.*** 2007,***E90-B***(11):3170-3180.

10. Seo H, Lee C: ***A NEW GA-BASED RESOURCE ALLOCATION SCHEME FOR A READER-TO-READER INTERFERENCE PROBLEM IN RFID SYSTEMS.*** In Paper presented in 2010 IEEE international conference on communications (ICC). Cape Town, South Africa; 23–27 May 2010.

11. International Organization for Standardization: ISO/IEC 18000-6C:2004: information technology – radio frequency identification for item management-Part 6: parameters for air interface communications at 860 MHz to 960 MHz. Geneva, Switzerland: International Organization for Standardization,; 2010.

12. International Organization for Standardization: ISO/IEC 18000-7:2008: information technology – radio frequency identification for item management – Part 7: parameters for active air interface communications at 433 MHz. Geneva, Switzerland: International Organization for Standardization; 2010.

13. EPC Global Inc.: EPC Radio-Frequency Identity Protocols Class-1 Generation-2 UHF RFID Protocol for Communications at 860 MHz–960 MHz Version 1.2.0. Brussels Belgium: EPC Global Inc.; 2008.

14. Abramson N: ***PACKET SWITCHING WITH SATELLITES.*** In Proc. National Computer Conference. ACM Press, New York; 1973:695-702.

15. Bueno-Delgado MV, Vales-Alonso J, González-Castaño FJ: ***ANALYSIS OF DFSA ANTI-COLLISION PROTOCOLS IN PASSIVE RFID ENVIRONMENTS.*** In Paper presented in proceedings of the 35th international conference of the IEEE Industrial Electronics Society. Porto,Portugal; 3–5November 2009.

16. Vales-Alonso J, Bueno-Delgado MV, Egea-López E, Alcaraz J, González-Castaño FJ: ***MULTI FRAME MAXIMUM LIKELIHOOD TAG ESTIMATION FOR RFID ANTI-COLLISION PROTOCOLS. IEEE TRANS. ON INDUSTRIAL INFORMATICS.*** 2011,***7***(3):487-496.

17. Milenkovic O, Compton KJ: ***PROBABILISTIC TRANSFORMS FOR COMBINATORIAL URN MODELS. COMB. PROBAB.COMPUT.*** 2004,***13***(4-5):645-675.

4

CHAPTER

DESIGN OF A HUMIDITY SENSOR TAG FOR PASSIVE WIRELESS APPLICATIONS

Xiang Wu [1], Fangming Deng [1,*], Yong Hao [2], Zhihui Fu [1] and Lihua Zhang [1]

[1] *School of Electrical and Electronic Engineering, East China JiaoTong University, Nanchang 330013, China*

[2] *School of Mechatronic Engineering, East China JiaoTong University, Nanchang 330013, China*

ABSTRACT

This paper presents a wireless humidity sensor tag for low-cost and low-power applications. The proposed humidity sensor tag, based on radio frequency identification (RFID) technology, was fabricated in a standard 0.18 μm complementary metal oxide semiconductor (CMOS) process. The top metal layer was deposited to form the interdigitated electrodes, which were then filled with polyimide as the humidity sensing layer. A two-stage rectifier adopts a dynamic bias-voltage generator to boost the effective gate-source voltage of the switches in differential-drive architecture, resulting in a flat power conversion efficiency curve. The capacitive sensor interface, based on phase-locked loop (PLL) theory, employs a simple architecture and can work with 0.5 V supply voltage. The measurement results show that humidity sensor tag achieves excellent linearity, hysteresis and stability performance. The total power-dissipation of the sensor tag is 2.5 μW, resulting in a maximum operating distance of 23 m under 4 W of radiation power of the RFID reader.

Keywords: humidity sensor; complementary metal oxide semiconductor (CMOS); radio frequency identification (RFID); rectifier; sensor interface

1. INTRODUCTION

Humidity measurement is essential for a wide range of applications in many fields including meteorology, agriculture, industrial control, medical instruments, etc. Humidity sensors usually measure relative humidity (RH) rather than absolute humidity. Relative humidity is the ratio of the moisture level to the saturated moisture level at the same temperature and pressure and expressed as a percentage. The main mechanisms to sense and measure RH include detecting changes in the optical [1,2], mechanical [3,4] and electrical [5,6] properties of sensing materials. The majority of the electrical humidity sensors in commercial use are of the capacitive type, which can offer lower power consumption and a less complex interface circuit compared with other types. Due to its good moisture absorption and compatibility with integrated circuit (IC) fabrication technologies, polyimide is a good candidate for moisture sensing films in capacitive humidity sensors for both micro-electro-mechanical system (MEMS) and complementary metal oxide semiconductor (CMOS) technologies [7]. Fabrication of humidity sensors in CMOS technology allows them to be easily integrated with other signal- processing circuits on a single chip, resulting in advantages including improved accuracy, reduced size and lower fabrication cost.

Radio frequency identification (RFID), as a wireless automatic identification technology, is widely applied in traffic management, logistics transportation, medicine management, food production, etc. [8]. Passive RFID tags offer several advantages such as battery-less operation, wireless communication capability, high flexibility, low cost and fast deployment, which all result in their extensive use in commercial applications [9]. The passive RFID tag collects the radiation energy from the RFID reader as its power supply. Hence, the power dissipation of the passive RFID tag, which determines the maximum operating distance of the tag, is crucial for the design of a passive RFID tag. Recently, with the rapid development of the Internet of Things and sensor technology, research on adding sensing functionality to RFID tag is become a hot topic [10,11,12]. This smart RFID sensing tag not only extends the application field of RFID, but also contributes to reduce the fabrication cost of RFID systems.

Previous research on capacitive CMOS humidity sensors was mainly focused on the humidity sensor element [13,14,15,16,17]. However the designs in [13,14,15,16] require some post-processing steps, the design in [17] uses materials and steps not commonly found in standard fabrication processes, which all resulting in increased fabrication costs. Benefiting from advantages such as lower packaging cost, smaller parasitic capacitance, smaller chip area, etc., fully integrated humidity sensors were recently introduced [18,19,20]. The designs in [18,19] reported humidity sensors with integrated interfaces. The design in [20] reported a wireless humidity sensor, but its operating frequency is 13.56 MHz which is only suitable for low speed and short distance applications.

Our motivation in this work has been to develop a wireless humidity sensor operating at 900 MHz frequency. For low-cost applications, the humidity sensor element is incorporated with the wireless transceiver blocks and the humidity sensor element is fabricated in standard CMOS technology without any post-processing. An ultra-low power sensor interface is introduced to ensure that the wireless sensor can work in passive mode. The rest of the paper is organized as follows: Section 2 presents the system architecture of the wireless humidity sensor and then the analysis of the design of key blocks is detailed in Section 3. Section 4 illustrates the measurement results and then conclusions are presented in Section 5.

2. SYSTEM DESIGN

The operating frequency plays a very important role in RFID systems. In general, the operating frequency defines the data rate of the system and the tag size. Ultra-high frequency (UHF) RFID systems are a better solution for next generation auto-ID applications because they have advantages including fast transmission speed, long communication distance and simple modulation schemes. The theoretical practicable operating power of an RFID tag P_t is calculated from the Friis transmission equation [21]:

$$P_t = E_r \bullet G_a \bullet \eta_r \bullet \left(\frac{\lambda}{4\pi d} \right)^2 \tag{1}$$

where E_r is the effective isotropic radiation power of a reader, G_a is the tag antenna gain, η_r is the RF-to-DC power conversion efficiency of the rectifier, and d is the communication distance. From this equation, it can be concluded that in order to achieve a longer communication distance, lower tag power and higher η_r of the rectifier are critical for the UHF RFID tag design because E_r is limited by regional regulations (4 W is the maximum transmitted power) and G_a is roughly determined by the allowable antenna area (1.64 for the $\lambda/2$ dipole antenna).

Figure 1 shows the architecture of the proposed passive wireless humidity sensor. The blocks shown, except for the antenna and matching network, are integrated on a single tag chip. The tag antenna, which is matched with the tag chip through the matching network, receives the electromagnetic waves from the RFID reader. The rectifier multiplies and transfers the received RF signal to a DC supply voltage for the subsequent circuitry. Once the output of the rectifier reaches the operating voltage, the power-on-reset (POR) block generates a reset signal for the sensor interface. Due to the good immunity to voltage supply, the regulator block in the conventional RFID tag architecture is not employed in this proposed architecture, which can be concluded in Section 3.3. The demodulator block is also not included in this architecture, which means that this wireless sensor will operate without any addressing as long as the sensor tag receives enough large signals from the RFID reader [22].

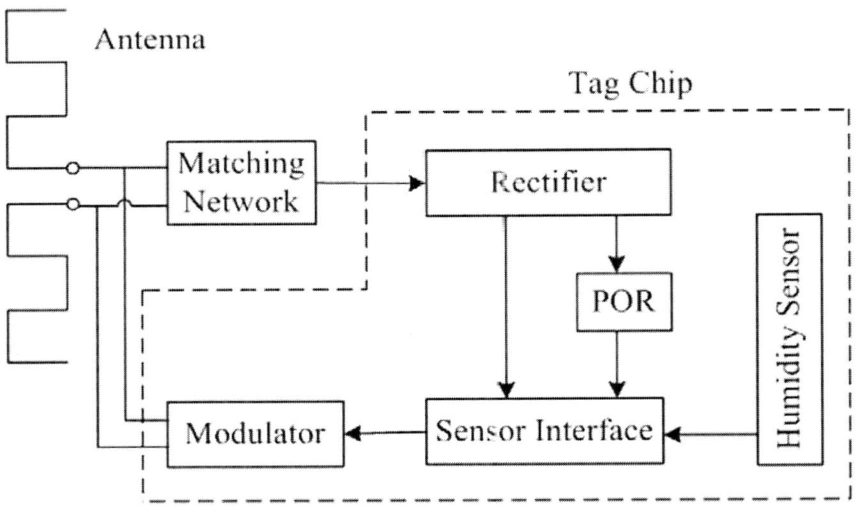

Figure 1. Architecture of the proposed wireless sensor.

For low-cost and small-size applications, the meandered dipole antenna is a natural choice for UHF RFID taga [23]. The modulator of the tag chip employs a low-power backscattering scheme, in which the RFID tag acts as a reflector that reflects the incident RF wave back to the RFID reader. Backscattering can be either amplitude shift keying (ASK) or phase shift keying (PSK) backscattering. ASK backscattering is much more simple and efficient than PSK backscattering [24], therefore this work employs an ASK backscattering scheme. Figure 2 shows the scheme of the ASK-backscattering modulation. If the tag intends to send digital "1" to RFID reader, the SW is opened which leads to the perfect match ($R_{in} = R_{ant}$) and then the tag entirely absorbs the electromagnetic wave from RFID reader. If the tag intends to send a digital "0" to the RFID reader, the SW is closed which leads to the complete mismatch ($R_{in} = 0$) and then the tag entirely backscatters the electromagnetic signal to the reader.

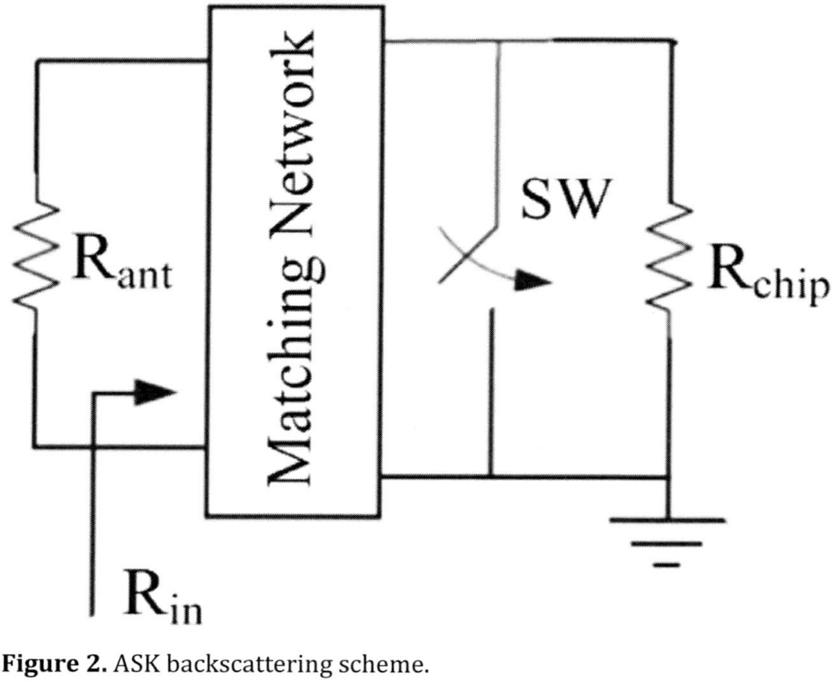

Figure 2. ASK backscattering scheme.

3. KEY BLOCKS DESIGN

3.1. Humidity Sensor

The reported CMOS humidity sensor designs [13,14,15,16] require some post-processing steps and the design [17] uses materials and steps not commonly found in standard fabrication processes, all of which will undoubtedly increase the fabrication cost. For a CMOS process, interdigitated top metal fingers, with polyimide filled into the finger gaps, can be utilized for capacitive humidity sensing [18,19,20]. The line-to-line coupling capacitance of the top metal is sensitive to the dielectric constant of the filling material. Furthermore, due to the high precision of photolithography, such a kind of sensing structure is highly reproducible with less inter-die variations.

Figure 3 illustrates the structure of the proposed capacitive humidity sensor in a standard CMOS process, resulting in the integration with other tag blocks on a single chip. The proposed humidity sensor was fabricated in the TSMC 0.18 μm 1P6M CMOS process. The top metal layer (Metal 6) was deposited and patterned with standard optical lithography and wet etching over the isolation layer to form the interdigitated structure. The sensing capacitor was then covered with a humidity-sensitive polyimide layer. As seen from Figure 3, L is the length of metal electrodes, S is the width of each electrode, and W is the distance between adjacent electrodes. The thickness of the polyimide layer is H, which is generally larger than metal thickness h. For an N finger array sensor, the total sensor capacitance can be expressed as [25]:

$$C_{sensor} = N\varepsilon_{wet}\frac{Lh}{W} \tag{2}$$

The metal thickness h and the electrode width S are limited by the CMOS process. The thickness of the polyimide layer H is a trade-off of between the sensitivity and response time. The selection of H has great influence on the sensitivity and response time of the sensor. Generally, a humidity sensor with a thicker polyimide layer has better sensitivity and longer response times. Considering the factors including sensor capacitance, chip area sensitivity and etc., this work chooses N = 40, L = 200 μm, W = 2.5 μm, S = 2.5 μm, h = 1 μm and H = 2 μm.

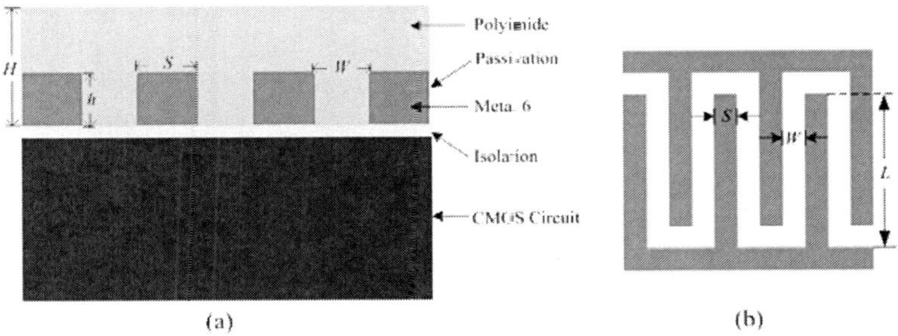

Figure 3. Proposed humidity sensor: (**a**) humidity sensor structure and (**b**) top view of the humidity sensor.

3.2. Rectifier

A popular performance metric of a rectifier is its power conversion efficiency which is defined as:

$$\eta_r = \frac{P_{out}}{P_{in}} \tag{3}$$

where P_{out} is the average DC output power generated at the output of the rectifier and P_{in} is the average RF power available at the input of the rectifier. From a circuit-level point of view, η_r is mainly degraded due to the forward drop of the switch diodes or transistors [26]. There are several techniques that are used to reduce the turn on voltage, including Schottky diodes, low V_{th} transistors and dual-poly floating gate transistors, which all require advanced CMOS processing at an additional cost [27,28]. In standard CMOS technology, a diode-connected MOS transistor is usually employed as switch transistor, but unfortunately the high threshold-voltage of MOS switch will greatly deteriorate the η_r of this architecture. The classic differential-drive architecture [29] is an excellent solution for high-efficiency and low-cost applications, however the η_r curve of this design drops rapidly on the two sides of the optimal point.

Inspired by Kamalinejad [30], this work employed a gate-boosting scheme to achieve a flat η_r curve. When the input power is smaller (or larger) than the optimal point of the η_r curve, an extra bias-voltage is added to positively (or negatively) boost the effective gate-source voltage of the switches. When input power is at the optimal point, the extra bias-voltage equals to zero. Thus, the bias-voltage is dynamic. According to the rules discussed above, Figure 4a shows the schematic of the proposed rectifier, which consists of two identical stage. The NMOS transistors M_{N11-22} and PMOS transistors M_{P11-22} form the differential-drive switch. The bias-voltage $V_{N1,2}$ and $V_{P1,2}$ are employed to boost the gate-source voltage of NMOS switch and PMOS switch, respectively. The large resistor R_S is added to block the AC component of $V_{N1,2}$ and $V_{P1,2}$. In CMOS process, R_S could be replaced by a PMOS transistor that operates in the cut-off region to avoid the large silicon area. Figure 4b shows the schematic of the dynamic bias generator circuit, which is directly supplied by the output voltage of the rectifier V_O. The transistors M_1-M_5 form the core of the reference voltage generator and the other transistors form the current mirrors. The required bias-voltage $V_{N1,2}$ and $V_{P1,2}$ can be achieved through by proper sizing of transistors.

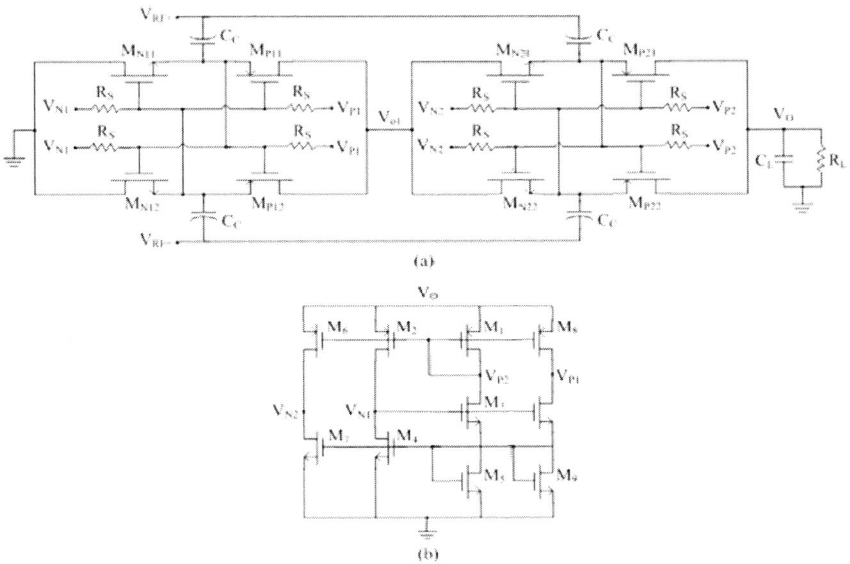

Figure 4. Schematic of the proposed rectifier: (**a**) two-stage rectifier and (**b**) bias-voltage generator.

3.3. Sensor Interface Design

The capacitive humidity sensor acts as a capacitor when it works with the sensor interface. There are several popular techniques to perform the capacitance-to-digital conversion function. The traditional conversion starts with a capacitance-to-voltage converter, which is then followed by a voltage-to-digital converter [31,32]. This technique can achieve high speed and high resolution performance, however, due to the use of an operational amplifier in switched-capacitor amplifiers (SCA), this technique results in too much power dissipation in the mW range. Accordingly, the inverter is proposed to replace the operational amplifier in SCA [18,33], resulting in a significant reduction of the entire power dissipation. Nevertheless, they still employ a relatively high supply voltage. Another conversion method in the time domain adopts pulse-width modulation for low-power applications [34,35]. This technique is suitably applied in the field of large-scale capacitance variation.

This work employs a phase-locked loop (PLL)-based architecture which can directly achieve the capacitance-to-digital conversion [36]. The proposed sensor interface is shown in Figure 5a. It consists of three blocks, including sensor-controlled oscillator (SCO), digital-controlled oscillator (DCO) and phase detector (PD). Both the SCO and the DCO are implemented as three-stage inverter-based ring oscillators. The sensor capacitor C_s acts as the variable load on a single stage of the SCO, thereby generating a sensor-controlled frequency f_s. The DCO is steered by the PD output signal b_o, which is a representation of the phase difference between the SCO and the DCO. The variable capacitive load on a single stage of the DCO consists of two capacitors, C_o and C_m. The capacitor C_o, designed equal to the quiescent value of C_s, always connected to the DCO. But the capacitor C_m, designed slightly larger than the maximum variation of C_s, is swapped in or out of the DCO depending on the feedback from the PD. The PD is simply composed of a single-bit d-flip-flop. When the entire feedback loop is locked, the average digital frequency f_d will correspond to the sensor frequency f_s. Therefore, b_o represents the digital value of the sensor capacitance. This paper employs current-starved ring oscillators for the design of SCO and DCO. As shown in Figure 5b, M_1-M_6 form the 3-stage inverter-based ring oscillator, which current is constrained by the current mirror M_7-M_{12}. Although the extra transistors of M_7-M_{12} need a higher supply voltage, the current flowing through the inverters can be constrained much smaller, resulting in lower power-dissipation and higher temperature stability.

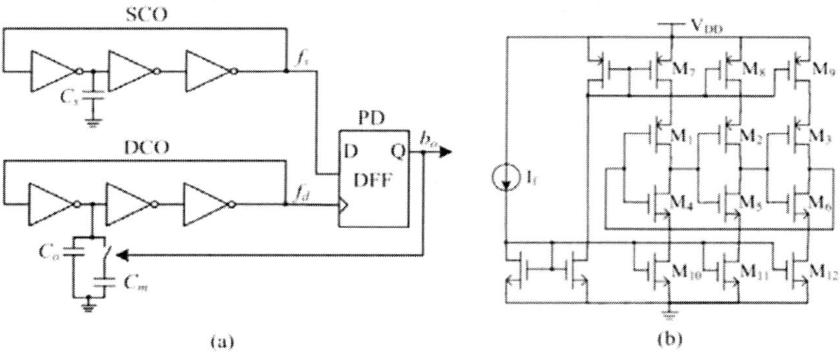

Figure 5. Proposed capacitive sensor interface: (**a**) architecture and (**b**) schematic of current-starved oscillator.

For this work, the humidity sensor's capacitance varies from 5 to 6.5 pF within the relative humidity range. Hence the capacitor C_o and C_m are selected as 5 pF and 2 pF respectively. Due to the fully digital architecture, the power supply of the interface is selected as 0.5 V. This architecture has an inherent immunity to supply voltage variations due to the fact that the digital output b_o is determined by the phase difference between the DCO and the SCO, resulting in the exclusion of a voltage regulator in the proposed wireless humidity sensor tag (seen in Figure 1).

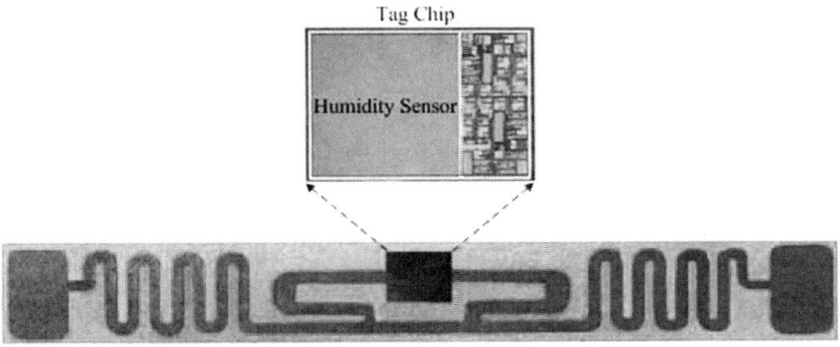

Figure 6. Photo of the proposed wireless humidity sensor.

Figure 7. Measurement environment.

4. MEASUREMENTS RESULT AND DISCUSSION

Figure 6 shows the proposed wireless humidity sensor, which was fabricated in the TSMC 0.18 μm CMOS process. The sensor tag equipped with a UHF antenna occupies an area of 80 × 6 mm^2 and the packaged sensor tag chip covers around 7 × 3 mm^2. The humidity sensor performance was measured in a Votsch VCL4003 temperature and humidity chamber. As shown in Figure 7, the wireless test environment consists of a special RFID test instrument, VISN-R1200 from VI Service Network and an anechoic box. The VISN-R1200 can work as an adjustable RFID reader and display the testing signals simultaneously. The RFID tag was tested in the anechoic box for the electromagnetic shielding.

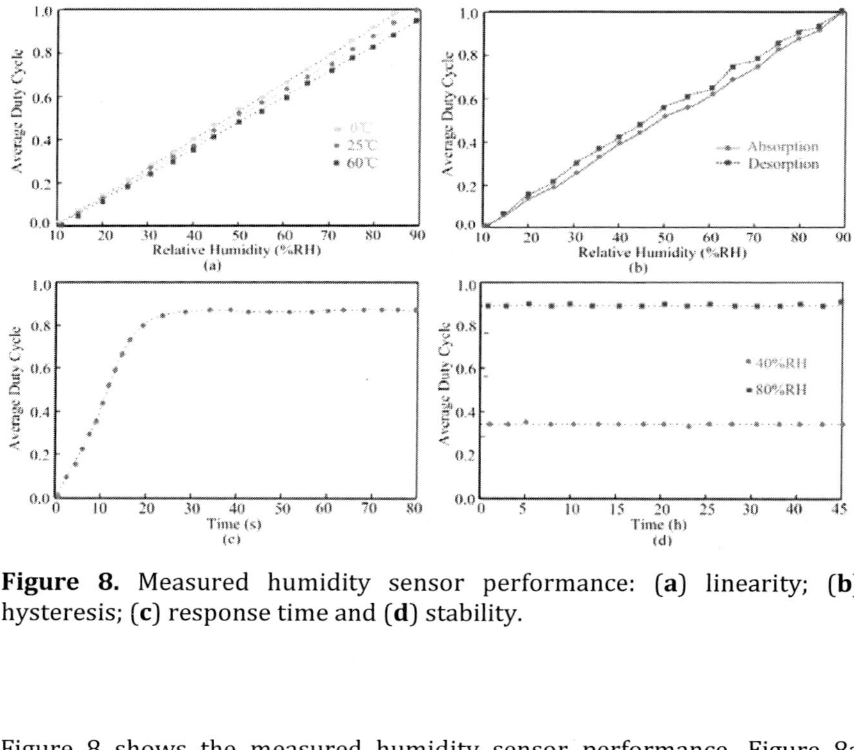

Figure 8. Measured humidity sensor performance: (**a**) linearity; (**b**) hysteresis; (**c**) response time and (**d**) stability.

Figure 8 shows the measured humidity sensor performance. Figure 8a illustrates the average duty cycle responses of the interface output with respect to the relative humidity at different temperature.

Within the relative humidity range from 10% to 90% RH, this humidity sensor achieves highly linear performance. Due to the temperature dependence of the dielectric constant of polyimide, the average duty cycle shows an acceptable maximum temperature variation of 10% from 0 to 60 °C. The humidity sensor achieves a sensitivity of 18.75 fF/%RH at 25 °C. The hysteresis performance of the sensor at 25 °C is shown in Figure 8b. The maximum difference between the moisture absorption and desorption at the point 55% RH is not exceeding 8%. Figure 8c shows that the response time of the humidity sensor is 20 s, which was measured to 90% point of the final steady state after an abrupt relative humidity change from 10% RH to 80% RH at 25 °C. The proposed humidity sensor was then tested for 45 h at the humidity of 40% RH and 80% RH respectively. As shown in Figure 8d, the measurement results showed excellent stability and no obvious drift was observed. Table 1 compares the proposed humidity sensor with previous designs. Our work achieves an average sensitivity, but it is fabricated without post-processing steps and integrates wireless transceiver blocks.

Table 1. Comparison of integrated humidity sensors.

Design	Sensor Structure	Process	Sensitivity	Fabrication Post-Processing	On-Chip Readout Circuit
[13]	Interdigitated	3 μm	5 fF/%RH	Yes	No
[15]	Parallel Plate	0.5 μm	303 fF/%RH	Yes	No
[16]	Interdigitated	0.35 μm	0.11 MHz/%RH	Yes	No
[17]	Woven Mesh	0.15 μm	1.78 mV/%RH	No	Yes
[18]	Interdigitated	0.16 μm	7.43 fF/%RH	No	Yes
[20]	Interdigitated	0.6 μm	30 fF/%RH	No	Yes
This work	Interdigitated	0.18 μm	18.75 fF/%RH	No	Yes

For a 900 MHz input frequency and 50 kΩ load, the η_r curve of the proposed rectifier is shown in Figure 9a. As compared to the conventional differential-drive rectifier [29], the proposed rectifier shows a flat η_r curve. When input power is 13 dBm, the two curves both reach the optimal point 69%.

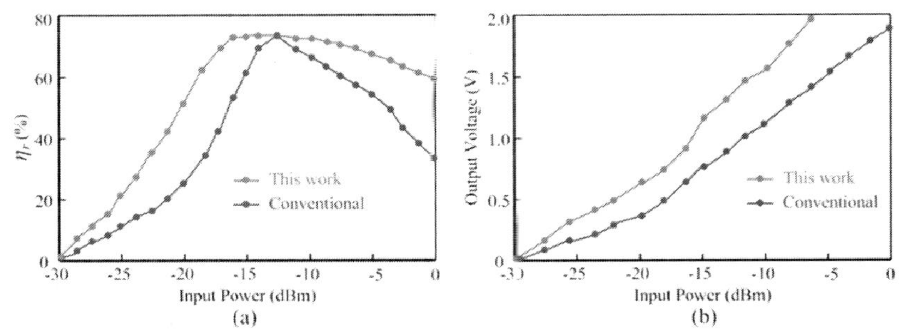

Figure 9. Performance comparison between conventional rectifier and this work: (**a**) power conversion efficiency and (**b**) output voltage.

As for the input power range which η_r is above 60%, the proposed rectifier and the conventional rectifier achieve −19 dBm and −7 dBm range, respectively. Figure 9b compares the output voltage of the proposed rectifier with that of the conventional rectifier for a 50 kΩ load at 900 MHz input frequency. The proposed scheme provides an output voltage of 0.6 V at the low input power of −20 dBm, while this output for the conventional counterpart is 0.3 V which is insufficient for many RFID applications.

The sensor interface performance was measured under a constant 0.5 V supply voltage from the rectifier. The performance comparison with previous designs is shown in Table 2. Owing to its simple architecture, the proposed sensor interface covers a reduced chip area and can operate on a 0.5 V ultra-low supply voltage. Despite the moderate effective number of bits (ENOB), this interface reduces power consumption significantly in contrast with the previous designs, making it especially suitable for passive RFID applications.

Table 2. Performance comparison of capacitive sensor interfaces.

Interface	Process (μm)	Supply (V)	Area (mm²)	ENOB (bits)	FOM (pJ/conv)	Power (μW)
[18]	0.16	1.2	0.15	12.5	8300	10.3
[33]	0.09	1.0	N/A	10.4	1.4	3.0
[34]	0.35	3.0	0.09	9.3	3.4	54.0
[35]	0.32	3.0	0.52	9.8	4.5	84.0
This work	0.18	0.5	0.01	6.8	1.6	1.1

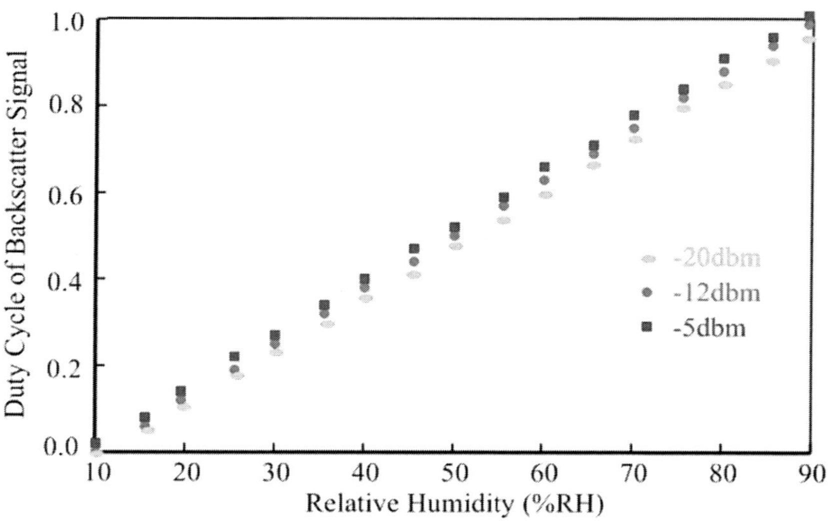

Figure 10. Duty cycle of backscatter signal versus relative humidity.

The measured minimum power dissipation of the wireless humidity sensor tag is 2.5 μW, resulting in a maximum operating distance of 23 m under 4 W RFID reader radiation power conditions. Finally, the measured backscattering performance of the sensor tag is shown in Figure 10. The measurement as repeated when the input power is −20 dbm, −12 dbm and −5 dbm, respectively. Due to the interface's immunity to the supply voltage, the three measured results show excellent linearity and consistency.

5. CONCLUSIONS

We have developed a humidity sensor tag for wireless applications. This wireless humidity sensor tag operates in the UHF band and employs an ASK modulator. The humidity sensor element is designed in standard CMOS technology without any post-processing, which results in integration with other tag blocks and low fabrication cost. A two-stage rectifier adopts a dynamic bias-voltage generator to boost the effective gate-source voltage of the switches in differential-drive architecture, resulting in a flat power conversion efficiency curve. The sensor interface, based on phase-locked loop theory, employs a fully-digital architecture. Despite the moderate ENOB, this interface can operate on ultra-low supply voltage and then reduces the power dissipation significantly compared to previous designs. The total power-dissipation of the sensor tag is 2.5 μW, resulting in a maximum operating distance of 23 m under 4 W RFID reader radiation power conditions.

ACKNOWLEDGMENTS

This work was supported by the National Natural Science Foundation of China (21265006) and the East China JiaoTong University Foundation (14DQ08).

AUTHOR CONTRIBUTIONS

Fangming Deng provided the instructions for system design for this research. Yong Hao helped perform the sensor measurements. Lihua Zhang and Zhihui Fu provided help in designing the humidity sensor and interface, respectively. All authors participated in the revisions of this manuscript.

CONFLICTS OF INTEREST

The authors declare no conflict of interest.

REFERENCES

1. Gu, B.B.; Yin, M.J.; Zhang, A.P.; Qian, J.W.; He, S.L. Optical fiber relative humidity sensor based on FBG incorporated thin-core fiber modal interferometer. Opt. Express **2011**, 19, 4140–4146. [

2. Correia, S.F.; Antunes, P.; Pecoraro, E.; Lima, P.P.; Varum, H.; Carlos, L.D.; André, P.S. Optical fiber relative humidity sensor based on a FBG with a Di-ureasil coating. Sensors **2012**, 12, 8847–8860.

3. Hong, H.S.; Chung, G.S. Humidity sensing characteristics of Ga-doped zinc oxide film grown on a polycrystalline AlN thin film based on a surface acoustic wave. Sensor Actuators B Chem. **2010**, 150, 681–685.

4. Wang, W.; Xie, X.; He, S. Optimal design of a polyaniline-coated surface acoustic wave based humidity sensor. Sensors **2013**, 13, 16816–16828.

5. Zampetti, E.; Pantalei, S.; Pecora, A.; Valletta, A.; Maiolo, L.; Minotti, A.; Bearzotti, A. Design and optimization of an ultra thin flexible capacitive humidity sensor. Sens. Actuators B Chem. **2009**, 143, 302–307.

6. Okcan, B.; Akin, T. A low-power robust humidity sensor in a standard CMOS process. IEEE Trans. Electron. Devices **2007**, 54, 3071–3078.

7. Boltshauser, T.; Azeredo, L.C.; Baltes, H. High sensitivity CMOS humidity sensors with on-chip absolute capacitance measurement system. Sens. Actuators B Chem. **1993**, 15, 75–80.

8. Li, B.; He, Y.G.; Guo, F.; Zuo, L. A novel localization algorithm based on isomap and partial least squares for wireless sensor networks. IEEE Trans. Instrum. Meas. **2013**, 62, 304–314.

9. Li, B.; He, Y.G.; Zuo, L.; Long, Y. Metrics of the application environment impact to the passive UHF RFID system. IEEE Trans. Instrum. Meas. **2014**, 63, 2387–2395.

10. Abad, E.; Palacio, F.; Nuin, M.; Zárate, A.; Juarros, A.; Gómez, J.M.; Marco, S. RFID smart tag for traceability and cold chain monitoring of foods: Demonstration in an intercontinental fresh fish logistic chain. J. Food Eng. **2009**, 93, 394–399.

11. Oprea, A.; Courbat, J.; Bârsan, N.; Briand, D.; de Rooij, N.F.; Weimar, U. Temperature, humidity and gas sensors integrated on plastic foil for low power applications. Sens. Actuators B Chem. **2009**, 140, 227–232.

12. Beriain, A.; Rebollo, I.; Fernandez, I.; Sevillano, J.F.; Berenguer, R. A passive UHF RFID pressure sensor tag with a 7.27 bit and 5.47 pj capacitive sensor interface. In Proceedings of the 2012 IEEE International Microwave Symposium Digest (MTT), Montreal, PQ, Canada, 17–22 June 2012; pp. 1–3.

13. Gu, L.; Huang, Q.A.; Qin, M. A novel capacitive-type humidity sensor using CMOS fabrication technology. Sens. Actuators B Chem. **2004**, 99, 491–498.

14. Zhao, C.L.; Qin, M.; Huang, Q.A. A fully packaged CMOS interdigital capacitive humidity sensor with polysilicon heaters. IEEE Sens. J. **2011**, 11, 2986–2992.

15. Kim, J.H.; Hong, S.M.; Lee, J.S.; Moon, B.M.; Kim, K. High sensitivity capacitive humidity sensor with a novel polyimide design fabricated by MEMS technology. In Proceeding of the 4th IEEE International Conference on Nano/Micro Engineered and Molecular Systems (NEMS), Shenzhen, China, 5–8 January 2009; pp. 703–706.

16. Dai, C.L.; Lu, D.H. Fabrication of a micro humidity sensor with polypyrrole using the CMOS process. In Proceedings of the 5th IEEE International Conference on Nano/Micro Engineered and Molecular Systems, Nice, France, 20–23 January 2010; pp. 110–113.

17. Nizhnik, O.; Higuchi, K.; Maenaka, K. Self-calibrated humidity sensor in CMOS without post-processing. Sensors **2012**, 12, 226–232.

18. Tan, Z.; Daamen, R.; Humbert, A.; Ponomarev, Y.V.; Chae, Y.; Pertijs, M.A. A 1.2-V 8.3-nJ CMOS Humidity Sensor for RFID Applications. IEEE J. Solid State Circuits **2013**, 48, 2469–2477.

19. Deng, F.M.; He, Y.G.; Zhang, C.L.; Feng, W. A CMOS humidity sensor for passive RFID sensing applications. Sensors **2014**, 14, 8728–8739.

20. Cirmirakis, D.; Demosthenous, A.; Saeidi, N.; Donaldson, N. Humidity-to-frequency sensor in CMOS technology with wireless readout. IEEE Sens. J. **2013**, 13, 900–908.

21. Wei, P.; Che, W.; Bi, Z.; Wei, C.; Na, Y.; Li, Q.; Min, H. High-efficiency differential RF front-end for a Gen2 RFID tag. IEEE Trans. Circuits Syst. II **2011**, 4, 189–194.

22. Kapucu, K.; Dehollain, C. A passive UHF RFID system with a low-power capacitive sensor interface. In Proceedings of the 2014 IEEE RFID Technology and Applications Conference (RFID-TA), Tampere, Finland, 8–9 September 2014; pp. 301–305.

23. Rao, K.V.S.; Nikitin, P.V.; Lam, S.F. Antenna design for UHF RFID tags: A review and a practical application. IEEE Trans. Antennas Propag. **2005**, 53, 3870–3876.

24. Ashry, A.; Sharaf, K.; Ibrahim, M. A compact low-power UHF RFID tag. Microelectron. J. **2009**, 40, 1504–1513.

25. Dai, C.L. A capacitive humidity sensor integrated with micro heater and ring oscillator circuit fabricated by CMOS–MEMS technique. Sens. Actuators B Chem. **2007**, 122, 375–380.

26. Kamalinejad, P.; Keikhosravy, K.; Mirabbasi, S.; Leung, V. An efficiency enhancement technique for CMOS rectifiers with low start-up voltage for UHF RFID tags. In Proceedings of the International Green Computing Conference (IGCC), Dallas, TX, USA, 3–5 November 2013; pp. 1–6.

27. Theilmann, P.T.; Presti, C.D.; Kelly, D.; Asbeck, P.M. Near zero turn-on voltage high-efficiency UHF RFID rectifier in silicon-on-sapphire CMOS. In Proceedings of the IEEE Radio Frequency Integrated Circuits Symposium (RFIC), Anaheim, CA, USA, 23–25 May 2010; pp. 105–108.

28. Mandal, S.; Sarpeshkar, R. Low-power CMOS rectifier design for RFID applications. IEEE Trans. Circuits Syst. I **2007**, 54, 1177–1188.

29. Kotani, K.; Sasaki, A.; Ito, T. High-efficiency differential-drive CMOS rectifier for UHF RFIDs. IEEE J. Solid State Circuits **2009**, 44, 3011–3018.

30. Kamalinejad, P.; Keikhosravy, K.; Mirabbasi, S. A CMOS rectifier with an extended high-efficiency region of operation. In Proceedings of the 2014 IEEE RFID Technology and Applications Conference (RFID-TA), Johor Bahru, Malaysia, 4–5 September 2013; pp. 1–6.

31. Paavola, M.; Kamarainen, M.; Laulainen, E.; Saukoski, M.; Koskinen, L.; Kosunen, M.; Halonen, K.A. A micropower-based interface ASIC for a capacitive 3-axis micro-accelerometer. IEEE J. Solid State Circuits **2009**, 44, 3193–3210.

32. Shin, D.Y.; Lee, H.; Kim, S. A delta–sigma interface circuit for capacitive sensors with an automatically calibrated zero point. IEEE Trans. Circuits System II **2011**, 58, 90–94.

33. Nguyen, T.T.; Hafliger, P. An energy efficient inverter based readout circuit for capacitive sensor. In Proceedings of the Biomedical Circuits and Systems Conference (BioCAS), Rotterdam, Netherlands, 1–2 November 2013; pp. 326–329.

34. Sheu, M.L.; Hsu, W.H.; Tsao, L.J. A capacitance-ratio-modulated current front-end circuit with pulsewidth modulation output for a capacitive sensor interface. IEEE Trans. Instrum. Meas. **2012**, 61, 447–455.

35. Tan, Z.; Shalmany, S.H.; Meijer, G.C.; Pertijs, M.A. An energy-efficient 15-bit capacitive-sensor interface based on period modulation. IEEE J. Solid State Circuits **2012**, 47, 1703–1711.

36. Danneels, H.; Piette, F.; De Smedt, V.; Dehaene, W.; Gielen, G. A novel PLL-based frequency-to-digital conversion mechanism for sensor interfaces. Sens. Actuators A Phys. **2011**, 172, 220–227.

5

CHAPTER

COGNITIVE RADIO FREQUENCY ASSIGNMENT WITH INTERFERENCE WEIGHTING AND CATEGORIZATION

Zeljko Tabakovic1* and Mislav Grgic2

1HAKOM, Croatian Regulatory Authority for Network Industries, Roberta F. Mihanovica 9, 10000 Zagreb, Croatia.

2Faculty of Electrical Engineering and Computing, University of Zagreb, Unska 3, 10000 Zagreb, Croatia.

ABSTRACT

Cognitive radio is one of the technologies promoting flexible and efficient use of radio frequency spectrum, thus solving spectrum scarcity problem. Frequency assignment is an integral part of the cognitive radio spectrum management and a critical point of success or failure of the cognitive radio concept. In this paper, cognitive radio frequency assignment with a novel interference weighting and categorization is proposed, as an extension of the solution to the graph coloring problem. In our approach, the edge weights quantifying the interference potential are appended to the conflict graph, co- and adjacent channel interference are treated, and dynamically changing local lists of blocked frequencies are included. We propose the improved cognitive radio saturation metric for the dynamic vertex ordering and to introduce interference categorization which will reduce the communication overhead. Using the proposed model, resource manager can quantify individual interference components, as well as aggregate interference from multiple users, resulting in more knowledgeable frequency decisions. Generalization of the proposed model is suggested. The suggested generalization consists of the selection of the central frequency and optimal bandwidth to be used, according to the user requirements. We have developed interference-sensitive algorithms for minimizing the interference and maximizing the throughput, both in centralized and distributed implementation. The results show significant reduction of the interference, improved spectrum efficiency, and increase in network throughput, comparing to the benchmark algorithms.

Keywords: Cognitive radio Frequency assignment Resource allocation Graph coloring Spectrum management Dynamic spectrum access Interference modeling

1 INTRODUCTION

The future market demand for the wireless broadband services, internet of things, machine to machine communications, and wireless data offload requires the deployment of the next generation wireless networks, which will need a rapid and a more flexible access to the radio frequency spectrum. Since radio frequencies are a limited natural resource, which cannot be saved for the future use or transferred from underused to overloaded areas, efficient usage of the frequencies has become one of the major concerns of the industry, governments, and scientific research. As a result of these research efforts, many different proposals for better and more efficient usage of the radio frequencies have emerged. One of the promising technologies, which could make a big paradigm shift towards the future wireless networks, is a cognitive radio (CR) technology. CR is introduced in [1, 2] and refers to intelligent radio communications system employing technology that allows the system to obtain knowledge of its operational environment, using technology to learn from the environment and dynamically adjust its operational parameters and protocols [3].

Problem addressed. CR frequency assignment represents the core of the cognitive radio process, since its unique features, along with reconfigurability and learning, make the cognitive radio what it is. The frequency assignment (FA) problem in wireless communications is well-studied, due to the extensive research of FA in mobile radio networks, wireless mesh networks, broadcasting networks, or in military applications [4–6]. Most of these studies do not consider primary-secondary spectrum sharing nor the dynamic availability of the spectrum due to the presence of primary users. Taking a deeper look into these research approaches, we can deduce that the cognitive radio FA has many unique and specific features, which makes CR unsuitable for transposition of the same FA models and algorithms from other types of wireless networks.

Task of the frequency assignment in the CR context is to select the best available radio frequency spectrum to satisfy user's communication needs, while avoiding causing harmful interference to the primary users (PUs) and minimizing influence to the other concurrent secondary users (SUs) sharing the same frequency band. Since the spectrum opportunities used for dynamic spectrum access in CR are limited in both time and space, CR has to be very fast and pragmatic in using the available spectrum bands. Suboptimal but fast converging and robust solutions are preferred in relation to the optimal one which usually requires extensive data sets and too much time to obtain results. Having this in mind, we can present basic characteristics of FA in CR:

- CR has to operate as a secondary service on non-interfering, non-protection basis, alongside other services using the same frequency band.
- CR has to work non-intrusively, has to protect PUs, and vacate the spectrum in case of a PU appearance.
- CR FA has to be fast, adaptive and easy to implement and does not necessarily have to provide an optimal FA.
- CR FA has to re-initiate and re-converge in cases of the quality degradation due to a dynamic changes in the radio channel or SU and PU activation.
- CR FA has to allocate the channels and also the spectrum fragments of a different bandwidth.
- CR FA has to work with the limited information in an environment with either cooperative or selfish SUs.
- CR FA has to be flexible, applicable in centralized and distributed manner.
- CR FA has to assign frequencies continuously and sequentially when a part of the PUs and SUs are already operating.
- CR FA has to be applicable in the heterogeneous wireless environment with the different classes of the dynamic spectrum access, different user requirements, and different protection requirements.

Our contributions. Our paper is focusing on the FA component of the dynamic spectrum management in the CR context. The problem of quantifying the interference using interference weighting and categorization in CR FA process, as well as CR FA algorithms with a numerical measure of interference levels between SUs as described in this paper, have not been previously addressed to the best of our knowledge. The FA in the CR networks presented in this paper is formulated as a variation of the graph coloring problem with the dynamic vertex ordering, where specific CR features are included. In this paper, the special attention is granted to the modeling and control of the interference. Numerical quantification of the interference is enabled with the inclusion of the conflict graph edge weights representing co- and adjacent channel interference between the SUs. We have developed the centralized and distributed algorithms with interference minimization and throughput maximization as objectives. Specifically, the contributions and novelty of this paper can be summarized in the following:

- The extension of the CR network conflict graph with:
    - Introduction of the continuous value edge weights quantifying the potential level of co- and adjacent channel interference between SUs
    - Incorporation of the influence of the adjacent channel interference by introducing an additional layer in the conflict graph
- Introduction of the SU interference categorization for interference susceptibility of the FA algorithms, while reducing the communication overhead
- The proposal of the spectrum fragments assignment with the identification of the central frequency and optimal bandwidth for the CR transmissions
- Introduction of a new saturation metric for a dynamic sorting of the vertices in the process of the sequential FA, taking into consideration the channel limitations due to the PU transmissions as hard constraints and a level of the interference from the adjacent assigned CRs as soft constraints
- The proposal and evaluation of the interference sensitive FA algorithms with the objective of:
    - Minimal total CR network interference
    - Maximal CR network throughput with consideration of the network interference as a comprehensive metric

In our paper, the condition of not causing harmful interference to the active PUs is a hard constraint, which has to be respected and cannot be violated by any FA plan. This is realized by the list of the blocked channels at each SU. Each SU has its own local list of the blocked channels, which is dynamically changed in time and space as the PUs frequency usage pattern constantly changes. The soft constraints are conditions which can be violated in the FA process if there is no better choice of the channel. The violations of the soft constraints are penalized by the FA algorithm, resulting in the deterioration of the objective function. The objective function takes into consideration the interference among SUs and the throughput achieved by the SU transmissions.

Our work differs from the previous works on the graph coloring-based CR networks FA in several aspects: the level of the interference between the SUs and between the SUs and PUs are numerically quantified using both the continuous and discrete weights; the influence of the co-channel and adjacent channel interference are considered; all individual interference contributions and cumulative interference values are monitored in the course of FA process. Proposed FA algorithms use this unique feature of the numerical quantification of the interference to precisely control and assign frequencies that have minimal interference influence in relation to other users. The advantages of the proposed interference sensitive FA strategy are numerous. It significantly reduces the interference in the CR networks, fully protects the PUs, reduces the frequency reuse distance between the CR users, improves spectrum usage efficiency, and increases the overall networks throughput with the available radio spectrum.

Paper organization. The rest of the paper is organized in the following way. Section 2 presents the overview of the previous research along with the comparison and main differences with respect to our paper. In Section 3, we formulate a CR network FA problem, present the system model and propose a novel framework for the interference sensitive FA in the CR. In Section 4, an interference estimation model used in the algorithm design is elaborated, characterization and categorization of co- and adjacent channel interference is introduced, and determination of weighting coefficients is elaborated. We describe the phases of the proposed FA algorithms, saturation metric, and coloring label with the objective of interference minimization and throughput maximization and the transmit power control strategy in Section 5. In Section 6, considerations of the centralized and distributed implementation of the FA algorithms are discussed. In Section 7, generalization to the frequency and bandwidth spectrum assignment are proposed, as well as implementation consideration in practical CR systems. In Section 8, we present an evaluation of the proposed algorithms' performance in the simulated environment of the PU and CR networks.

Finally, we conclude the paper and outline the key results of this paper in Section 9.

2 RELATED WORK

Spectrum management is technical, procedural, and policy approach to the planning, coordination, and managing the use of the electromagnetic spectrum as a limited resource. The CR spectrum management functions have some specific capabilities, since the CR networks are focusing on the secondary access and spectrum sharing, while protecting the PUs and minimizing interference with the other SUs. An insightful overview of the spectrum management in the CR networks is presented in [7, 8]. The problem of the FA is commonly solved using different heuristic methods, the graph theory, game theory, or linear programming. In the FA modeled as a the graph coloring problem, the task is to color all the nodes of the graph with the minimum number of colors, in a way that no two adjacent nodes (nodes connected with an edge) have the same color [4–6]. A comprehensive analysis of the frequency assignment in the CR networks can be found in [9, 10].

The comparisons of the different target objectives, the approaches for the FA, and the used methods for the selected previous works presented in this section are summarized in Table 1. A deeper discussion on the insights of the related work and the main differences with respect to this paper are presented below. In [11], the authors propose a spectrum decision framework determining a set of spectrum bands taking into account the application requirements and dynamic nature of spectrum bands. For the real time applications, authors propose spectrum decision with the objective of total capacity variance minimization, while for the best effort applications objective is network capacity maximization. The authors propose a new metric for calculating the capacity of a CR user, namely, the expected normalized capacity of a user taking into account network switching delay. This spectrum switching delay includes the time intervals for the spectrum decision process in the base station, signaling for establishing new channels and the RF front-end reconfiguration. The dynamic resource management for the spectrum decision is elaborated including admission and decision control. The authors propose centralized spectrum management using linear integer optimization method with a single channel assignment. Two phase CR network power control and channel allocation is proposed in [12]. In the first phase, distributed power updating process is employed in order to maximize the coverage of the CR network, while assuring that the signal-to-interference-plus-noise ratio (SINR) constraints at all PU receivers are met. In the second phase, a centralized channel allocation scheme based on weighted bipartite graph matching is used. The objective of the channel assignment is the maximization of the total network throughput while protecting the PUs. Both, uplink and downlink scenarios, are investigated with distributed and centralized implementation. The authors consider a cellular CR network with base and terminal stations, where the interference between the SUs are not considered. Contrary to this, in our approach, the SUs protect the PUs using the local list of blocked channels, the interference between the SUs is measured with conflict edge weights, and FA algorithms consider the interference quantification in the frequency decisions. In [13], the FA problem for the opportunistic spectrum access is investigated. The authors propose a novel FA approach using color-sensitive graph coloring. The authors present a solution of the graph coloring problem using the sequential coloring with a vertex labeling. The vertex labeling is based on the channel capacity divided by the vertex degree, giving priority to the most valuable vertices. Based on those premises, the authors propose coloring algorithms with an objective of maximal cumulative reward, maximal minimal reward, and maximal proportional fairness. Results of the simulations show improvement of the network performance in throughput and in interference reduction compared to the other algorithms. In [13], the authors use binary edge conflict graph and interference free channel assignment. In our proposal, the interference is continuous or have discrete characteristic values. The FA algorithms allow the interference with matching network penalty. In [14], localized efficient channel assignment in the CR networks is proposed with a goal to maximize node connectivity after the FA, which is important for packet delivery. Partition of the network into two-level trees, conflict resolution strategy and trimming of nonessential links are proposed. Two basic localized algorithms and an advanced localized algorithm for the FA are presented. In assigning, the priority is given to the nodes with a higher link degree. Node-link-based algorithm with Hungarian method of maximal matching has almost twice the broadcast reachable nodes and assigned link rate, compared to the basic algorithms. As opposed to our approach, in the proposed model, the CR throughput is not considered, all channels have the same performance, there is no transit power control, and the interference range is the same as the communication range. In [15], the authors consider cooperative cognitive radio networks and propose the resource allocation in an OFDMA-based CR network with the SUs relaying data of the PUs. A new flexible channel cooperation (FLEC) design of the PU and SU cooperation is proposed. In the first time slot, PUs transmit to SU. In the second time slot, the SUs transmit to the primary BS. A game theory Nash bargaining-based solution for relay and subchannel assignment, as well as strategy optimization and power control, is considered. Centralized and distributed algorithms are proposed and show throughput improvement comparing to the conventional channel cooperation model. Distributed bargaining algorithm converges within 20 iterations, while centralized algorithm proves to be impractical due to the slow convergence. Consequently, the authors propose a heuristic method for the centralized problem. In [16], the authors propose a graph coloring-based fair channel allocation for base stations self-coexistence in the CR networks. The proposed solution is different from our approach in the following: each base station has a list of the available channels with equal characteristics and throughput. Graph representation of the system employs a binary interference model between the base stations. The proposed model allows operation of the CR networks with the different traffic requirements and guarantees the certain quality of service (QoS). Depending on the access category and priorities for the different packets, quality metric and share of the resources is determined. Simulation results show fairness improvement in terms of bandwidth allocation ratio and faster convergence than traditional list multi-coloring algorithm. An opportunistic distributed channel graph coloring algorithms for the spectrum access in dynamic spectrum access networks is presented in [17]. Maximum spectrum packing algorithm uses the local topology information of one-hop neighbors assigning a channel to the node with the highest degree. The algorithm ensures that all the SUs obtain one color and some SU's multiple colors. The algorithm assigns the color to the node with the highest degree and also enables assigning multiple colors to some nodes, thus improving overall spectrum efficiency. The improved heuristic with probabilistic color selection using truncated geometric distribution based on node degree ranking reduces the algorithm termination time. In probabilistic heuristic, the SUs request different channels with the probability that depends on the relative degree of the SU and its one-hop neighbors, reducing the possibility of two users contend for the same frequency. Random backoff is used to resolve the contentions of the same degree users during channel access. The performance of the algorithm is tested on randomly generated Bernoulli graphs with specified graph density. Comparing to our approach, proposed model does not consider the protection of the PUs and only simple binary conflict graphs are considered without interference susceptibility in the FA. In [18], the authors consider downlink resource allocation for hybrid macro/femtocell LTE-advanced network. Both macro and femto base stations have the CR functionalities. The cross tier interference is reduced by using the cross-cognition scheme, where base stations have higher priority in utilizing its licensed spectrum and lower priority in opportunistically utilizing non-licensed spectrum. The largest degree first based graph coloring algorithm is applied for the frequency block assignment reducing intra-tier interference between the femtocell nodes. Simulation results show increased throughput of the CR femtocell network compared to non-spectrum shared macro-femto cell network and improved interference mitigation of the graph coloring resource allocation compared to the random resource allocation. The problem of the FA to the communication links in the multi-hop wireless mesh networks with the objective of minimizing overall network interference is addressed in [19]. In the proposed mesh network, router nodes have multiple radio interfaces. The authors propose a centralized algorithm based on the heuristic Tabu search technique, where in the first phase, random initial FA solution is improved using the iterative algorithm with Tabu list, and in second phase, the algorithm eliminates remaining interface constraints. Distributed algorithm is based on a greedy heuristic algorithm for Max K-cut problem of the constrained graph optimization, where algorithm chooses frequency which produces the largest decrease in the local interference. The authors also propose an extension of the model with consideration of the non-orthogonal channels and non-uniform node traffic useful for practical implementation. The empirical evaluation on randomly generated network graph shows that proposed algorithms perform close to lower bounds established by the semi-definite and integer linear programming. However, compared to the approach presented in this paper, the authors assume quasi-static assignment of the channels, all channels are not available for the FA to all users, binary interference model is assumed in simulations, and finally, there is no sharing between the primary and the secondary users. In [20], a game theoretic approach for the distributed channel selection and power allocation in the cognitive radio networks is investigated. Proposed algorithm enforces the cooperation between the SUs in order to reduce the energy consumption and to find Nash equilibrium. Game theory model fits well with the cognitive FA problem because the spectrum decision of one of the SU influences the performance of neighboring SUs. Potential game formulation and no-regret learning model have been presented as a solution to the channel allocation problem. The influence of the cooperation on observed SINR and power consumption is investigated. Results show that the proposed iterative algorithm converges to a pure strategy Nash equilibrium solution.

So far, all the work done on the CR networks graph coloring FA employed simplistic binary co-channel interference model in order to assign frequencies and optimize utility function [10, 12–14, 16–19]. As we know, the interference control and management are of utmost importance for the spectrum sharing and efficient spectrum management in the CR systems. The CR FA algorithms have to deal with the mutual interference from co-channel or adjacent channel transmissions, which may cause significant performance degradation. Nevertheless, in the existing work, it was not possible to assess, quantify, or compare interference influence of the CR frequency selection on the neighboring SUs, nor to share spectrum between users when the interference exists, but it is bearable. Also, the adjacent channel interference was mainly neglected. To alleviate this deficiency, we propose a novel conflict graph model and algorithms with interference weighting and categorization, which enable interference sensitive frequency decisions, better interference control, and quantitative estimation of the individual components, as well as the calculation of the cumulative interference at the neighboring users. As a result, the proposed FA in the CR is performed with the interference sensitivity, achieving the efficient spectrum utilization, while enabling reliable transmissions.

Table 1 Summary of FA problem solutions presented in related work

Ref.	Objective	Approach	Method
[11]	Throughput/ variance	Centralized single ch	Linear integer optimization
[12]	Power/total throughput	Distributed / centralized	Graph theory
[13]	Fairness/throughput	Centralized mult ch	Graph theory
[14]	Node connectivity/ interference	Local	Heuristic algorithm
[15]	Throughput surplus	Distributed / centralized	Nash bargaining / game theory
[16]	Fairness/QoS	Distributed list coloring	Graph theory
[17]	Fairness	Local multi ch	Graph theory / heuristic
[18]	Throughput	Macro BS femtocell	Graph coloring
[19]	Interference	Wireless mesh	Graph coloring
[20]	Interference/power	Distributed	Game theory

3 THE PROPOSED FA FRAMEWORK

The CR spectrum management task is to achieve efficient wireless communications without causing a harmful interference or service

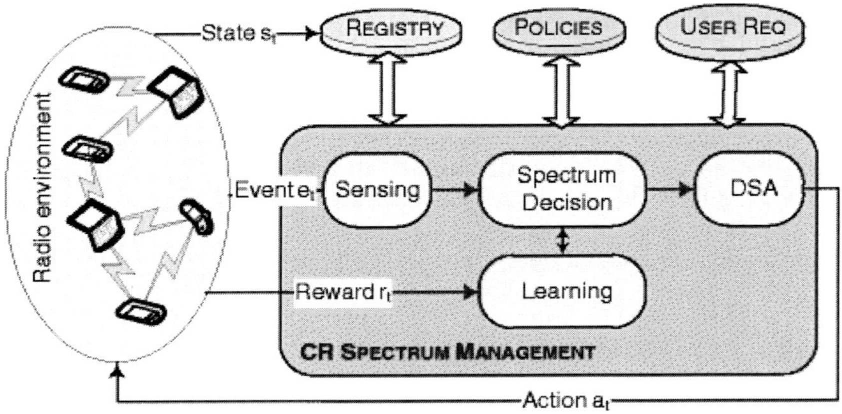

Figure 1 CR spectrum management framework

interruptions. The CR spectrum management functions have some specific capabilities since the CR networks are focusing on the secondary access and spectrum sharing, while protecting the PUs and minimizing interference to the other SUs. A functional overview of the CR spectrum management framework is shown in Fig. 1. It consists of four major steps:

- Spectrum sensing. A CR user can only utilize temporary unused parts of the spectrum. Therefore, CR should monitor the available spectrum bands, collect the information on the spectrum use, and identify possible spectrum holes and their characteristics.
- Spectrum decision. Based on the spectrum availability information acquired through the spectrum sensing, policy guidelines, user requirements, and registry of the spectrum use, CR characterizes radio frequency spectrum possible for various models of the dynamic spectrum access. In the spectrum decision, CR or the centralized entity determines the carrier frequency, channel bandwidth, transmission power, modulation, coding, communication technology, together with other operational, and technical parameters used for the CR operation.
- Dynamic spectrum access (DSA). The CR reconfigures its technical parameters in line with the selected operational technical parameters and operates in order to satisfy its primary goal of successful communication with a required QoS.
- Learning. Since CR is operating in heterogeneous radio environment with different user characteristics, different requirements and many parameters determining its environment and performance, CR has to

adapt to the constantly changing environment, observe performance of its operation, and has to adapt its spectrum decision function using reinforcement learning.

The process of assigning frequencies to the CR users as part of the spectrum decision step is of crucial importance for the CR network functioning and represents a focal point of the efficient radio spectrum use, distinguishing the CR from other wireless communications systems.

3.1 Problem formulation and system model

In this paper, we are focusing on the FA component of the CR spectrum management. We will interchangeably use the following terms: radio links and vertices, frequencies and colors, interference and edges, cognitive radio and secondary user. The problem of the FA in the CR networks can be informally described as follows: given a set of the transceivers with the CR capabilities, a network of the PU transceivers with the administrative licenses and priority access rights to the radio frequency spectrum, and predefined constrained number of the frequencies, we wish to efficiently assign radio frequencies to the CR users. The CR FA has to satisfy the following requirements:

- CR self-goal: successfully transmit as much information as possible with required QoS
- CR network goal:
    - Minimizing the total network interference
    - Maximizing the network throughput
- Requirements towards other users: not to cause excessive interference with the operating PUs, to keep the CR mutual interference under the reasonable limit and to efficiently share the radio frequency spectrum

The FA problem in the wireless communications can be modeled as a graph-theoretical problem by defining an undirected conflict graph describing the interference relations and assigning frequencies by coloring vertices. The graph coloring problem is known to be NP-complete problem [21] and therefore computationally hard. Although any given solution of the NP-complete problem can be verified in the polynomial time, there is no known efficient way to find the optimal solution. Also, the time required to solve the problem increases very quickly as the size of the problem grows.

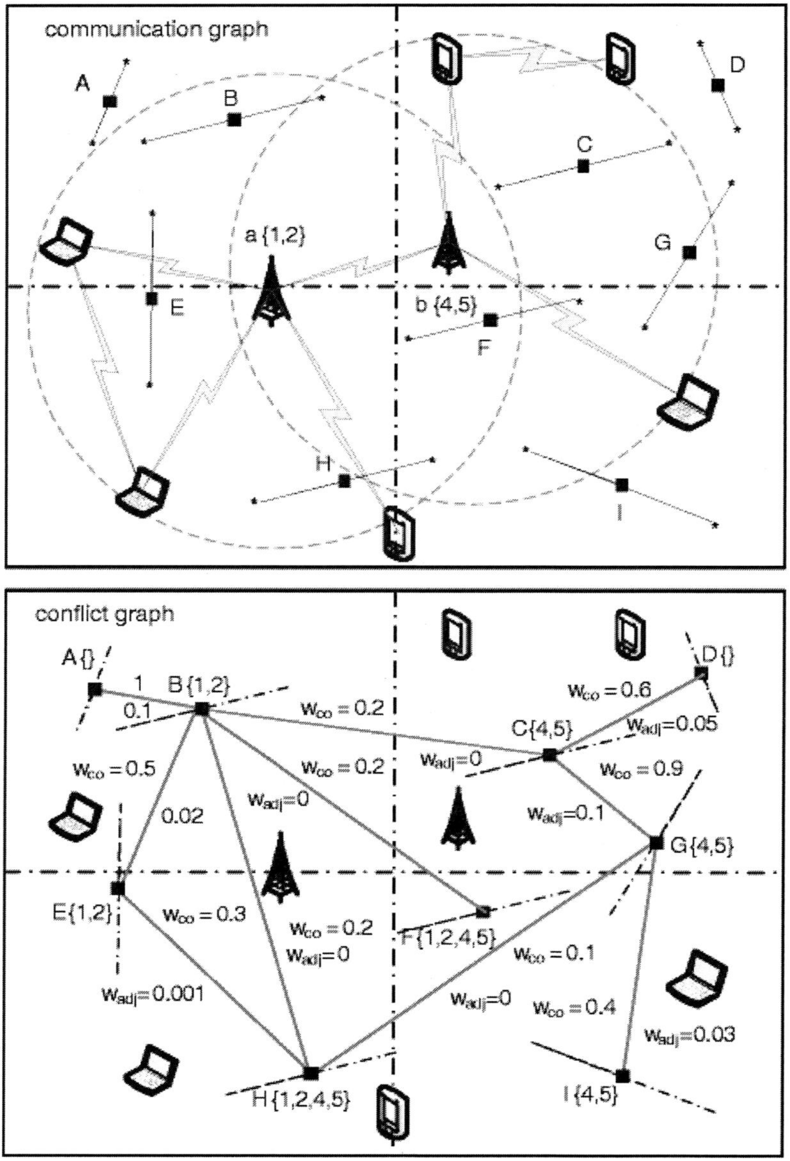

Figure 2. Communication and conflict graph illustration. The figure shows example of dynamic spectrum access scenario with corresponding communication and conflict graph

For example, 3.16 million years are needed to find a solution to the FA problem using brute force search approach for 20 terminals and 10 frequencies with 1 μs per solution [6]. Considering the basic characteristics of the FA in the CR already mentioned, and the computational complexity of finding an optimal solution, we are directing towards the use of the heuristic method for solving this problem. Although the heuristic algorithm leads to a suboptimal solution, it provides a reasonably good solution in the limited time, which works well in the practical applications of the assigning frequencies to the CR SUs.

In this paper, the CR FA is modeled as a graph coloring problem using novel dynamic vertex ordering, interference weighting, and categorization. To illustrate the concept presented in this paper, an example of the dynamic spectrum access scenario is depicted in Fig. 2, with the corresponding communication and conflict graph. The communication graph represents the communication links between the radio transceivers, while conflict graph shows possible interference between the radio links. In the communication graph, the PU and SU transceivers are represented as the vertices, while communication links between radio transceivers are represented by communication graph edges. In the communication graph shown in Fig. 2, PU network is illustrated as two base stations serving a set of PUs, establishing the primary links, while CR network deployed in the same area is represented by a set of CR links {*A*, *B*,...,*I*}. The circles illustrate the service zones of the PU networks. The corresponding conflict graph is shown in Fig. 2, where the vertices represent the CR communication links with a dynamically changing local list of blocked frequencies due to PU transmissions. In the same figure, the edges show possible interference occurrences where co- and adjacent channel interference levels are proportional to the edge weights coefficients.

Let us assume there are five available frequencies represented by numbers, which are opportunistically available to the CR operation. Since the PU and the CR networks share the same radio frequency band, the frequencies temporarily used by PUs cannot be utilized by the CR users if they are in the interference range, and therefore, those channels are blocked for corresponding CR link indicated by the corresponding blocked channels list. For the CR link, *B* channels {1,2} cannot be used due to the possible interference with the service area of the PU base station *a*, for the CR link *C* channels {3,4} are blocked, for the CR link *F* channels {1,2,4,5} are blocked, etc. The conflict edges represent the potential interference between the CR links with two edge weights per edge indicating co-channel and adjacent channel interference potential. In Fig. 2, the edge between the CR link *A* and CR link *B* has weights of 1 for co-channel and 0.1 for the adjacent channel, indicating very high interference potential between those links, and

Table 2 Important notation and acronyms

Notation	Description
$G_c = (V_c, E_{co}, E_{adj})$	Conflict graph with vertices V_c and edges E_{co}, E_{adj}
$F = \{1, .., M\}$	Set of M available frequencies
B_v	List of blocked frequencies at CRv due to PU hard constraints
$W_{co} = \left\{w^{co}_{vi\,vj}\right\} \in [0, 1]$	Set of co-channel continuous interference weight coefficients
$W_{adj} = \left\{w^{adj}_{vivj}\right\} \in [0, 1]$	Set of adjacent channel continuous interference weight coefficients
N_{PU}, N_{SU}	Number of primary and secondary users
$h^{PUi'}_{PUi}\ h^{SUk'}_{PUi}$	Channel coefficients between PU Tx i', SU Tx k', and PU Rx i
$h^{SUk'}_{SUl}\ h^{PUj'}_{SUl}$	Channel coefficients between SU Tx k', PU Tx j', and SU Rx l
$P_{Rx\_SUl\_D}, P_{Rx\_SUl\_I}$	Desired received signal strength and interfering received signal strength at the lth SU
$P_{Tx\_PUm'}, P_{Tx\_SUl'}$	Transmitting power at m'th PU and l'th SU
S_{vi}	Saturation metrics label
Col_{vi}	Vertex coloring argument

Acronyms	Description
CR	Cognitive radio
FA	Frequency assignment
PU, SU	Primary user, secondary user
Tx, Rx	Transmitter, receiver
QoS	Quality of service
CminSumInt	Centralized minimum cumulative network interference FA algorithm
CMaxSumCap	Centralized interference sensitive maximum throughput FA algorithm
DminInt	Distributed minimum interference FA algorithm
DMaxCap	Distributed interference sensitive maximum throughput FA algorithm

therefore, those links cannot use the same channel. The edge between the CR link C and CR link D has a conflict weight of 0.6/0.05 indicating a medium interference potential and the edge between CR link B and CR link F has a weight of 0.2/0 indicating a low co-channel interference and no adjacent channel interference. This indicates the possibility for sharing the channel. The existence of the conflict graph edges and the corresponding weights depends on the propagation channel characteristics and the antenna discrimination between the CR link pairs analyzed. For example, between the CR link A and CR link E, there is no edge because the terrain obstacle prevents possible interference path and corresponding links can use the same channel without danger of the mutual harmful interference. Practical implementation of constructing conflict graph and determining its weights is elaborated in Section 7.2.

In our work, we abstract the CR network topology into the communication graph $G=(V,E)$ as a general undirected graph over the set of network transceivers represented by the vertices V and PU or CR links represented by the edges E in Fig. 2. The FA channel selection constraints and potential interference between the radio links are represented by a conflict graph $G_c=(V_c,E_{co},E_{adj})$, where V_c is set of conflict graph vertices, and E_{co} and E_{adj} are co-channel and adjacent channel conflict edges with associated interference potential quantified by the edge weights w_{co} and w_{adj}. Using the described model, the FA problem can then be further formulated as a complete frequency assignment of the CR network represented by a conflict graph $G_c=(V_c,E_{co},E_{adj})$. The spectrum of interest is divided in M frequencies represented by constrained set of the radio frequencies $F=\{1,..,M\}$, and the dynamically changing local set of blocked frequencies protecting the PU transmissions represented by B_v. Using the abovementioned notation, frequency assignment is a graph coloring function $f:(G_c,B_v,W_{co},W_{adj})\rightarrow F$, such that the objective function is maximized. To facilitate further discussion, we will use the notation and acronyms as defined in Table 2.

3.2 Framework overview

The CR FA task is to assign the frequencies to the SUs when and where they request access to the spectrum, to assign new frequencies to the CR users when co- or adjacent channel PU transmission occurs in the interference range, or when the quality of service reduces below the requested threshold. As a result of that, we are facing with the task of assigning frequencies in a partially colored graph. In that respect, the FA is a continuous process, as it is not possible to assign frequencies and optimize coloring to the entire CR network at the same moment. There is a need to constantly assign and re-assign frequencies to active SUs. If the PU reactivates, new SU appears, or signal level drops due to dynamic characteristics of the radio channel, the FA process has to be re-initiated and channel allocation has to re-converge for the SUs influenced by these changes.

Based on the system model described in Section 3.1, we have developed a novel framework for assigning the frequencies in the CR environment. In the preparatory step, input data and the CR technical parameters are determined, including available frequencies, compatibility criteria, CR coverage area, and the interference potential. In the second step, the CR adjacency, interference levels, and weights are estimated, followed by communication and the conflict graph construction determining whether the CRs can share radio frequencies. In the CR FA step, an iterative process is performed consisting of labeling and ordering of the CRs, selecting the next transmitter to assign the frequency, and selecting the frequency maximizing network objective.

In the previous research of the CR FA application of graph coloring, the interference model assumed the conflict edges to be binary variables representing the interference or no-interference occurrence. The adjacent channel interference was not taken into account since the channels were assumed to be orthogonal. In the proposed FA framework, the CR network conflict graph illustrated in Fig. 2 is extended with the novel features:

- The local list of the blocked frequencies B_v associated with each conflict graph vertex representing frequencies which cannot be used by CR
- Two layers of edges in conflict graph taking into consideration co-channel interference E_{co} and adjacent channel interference E_{adj}
- Continuous interference weight coefficients w_{co} and w_{adj} associated to the conflict graph co-channel and adjacent channel edges incorporating quantification of possible interference between adjacent vertices corresponding to the interference "strength"
- The categorization of the interference weights reducing communication overhead

Since the CR user has to protect and not cause the interference to the licensed PUs, each CR keeps its own list of blocked frequencies and the FA algorithm avoids selecting the channels causing harmful interference to the PUs which are temporarily operating in the close vicinity of the observed CR. As the PUs frequency usage is constantly changing, the list of the blocked channels at the CR is dynamically updated using the spectrum sensing, the geo-location database or some other method of the spectrum awareness. In the proposed approach, the FA is interference sensitive, as the edges have associated co-channel and adjacent channel weights determining the level of the interference between the vertices. In this paper, we are proposing the algorithms which can be categorized under the sequential graph coloring class of algorithms with a saturation degree dynamic vertex ordering. This class of algorithms was first introduced by Brélaz [22] as saturation largest first algorithm known under the name Degree of saturation algorithm (DSATUR). In DSATUR algorithm, the vertices are ordered by a decreasing order of saturation, where the saturation degree is determined as a number of different colors to which a vertex is adjacent. In [13], color-sensitive graph coloring is used where dynamic vertex ordering is determined using labeling based on channel capacity divided by vertex degree and selecting most valuable SU to assign frequencies first (CR link contributing the most to the network throughput).

In our approach, we are also using labeling and dynamic SUs ordering to determine the prioritization of assigning frequencies to the CR users, but with the different saturation metric and network objectives compared to the existing literature on graph coloring FA. We are selecting the most difficult SUs to assign frequencies first using a specially designed CR saturation metric. Proposed novel CR saturation metric is determined as a level of freedom in selecting the frequency for the SU, taking into consideration channel limitations due to local PUs transmissions and a level of interference from adjacent assigned SUs as a combination of hard and soft frequency assignment constraints. The saturation metric is calculated using the interference weight coefficients. In this paper, we are proposing two different FA problem solutions with objective functions determined as minimal total network interference and as maximal throughput with the interference susceptibility.

Taking into account the CR network topology, we are considering centralized FA where centralized resource manager optimizes frequency use on the network or cluster level, and distributed FA where the selection of the frequencies is performed on a level of CR and its neighboring nodes. More detailed algorithm description and the mathematical formulation of the saturation metrics and objective functions are provided in Sections 5 and 6.

4 INTERFERENCE MODELING AND CHARACTERIZATION

4.1 Interference model

Modeling and estimation of the mutual interference are of vital importance for exploiting the potential of the radio spectrum band to the full extent, while keeping the interference under control and below level causing the network performance deterioration. Since the interference is the primary limiting factor influencing the performance of the CR networks, we have built our FA model with the interference susceptibility. The determined level of interference is a basis for the interference weighting and categorization used in the proposed FA algorithms. Channel model for the co-existing PUs and cognitive SUs is illustrated in Fig. 3. The figure shows one PU transmitter-receiver pair (PU Tx1' - PU Rx1) and two SU transmitter-receiver pairs (SU Tx1' - SU Rx1, SU Tx2' - SU Rx2). The received signal at Rx is noted by *Y*, transmitted signal from Tx is noted by *X* and *Z* is the noise. Lines between Tx and Rx represent TX-Rx channel paths, where each Tx is connected to all receivers with corresponding channel coefficients *h*. We assume that the quasi-static fading is present, and the channel coefficients between users are considered to be independent random variables. Channel coefficients take into consideration the total loss of the radio link between the transmitter and the receiver, as well as the antenna gain and the propagation effects (e.g., large scale fading due to the shadowing and small scale fading due to the multipath).

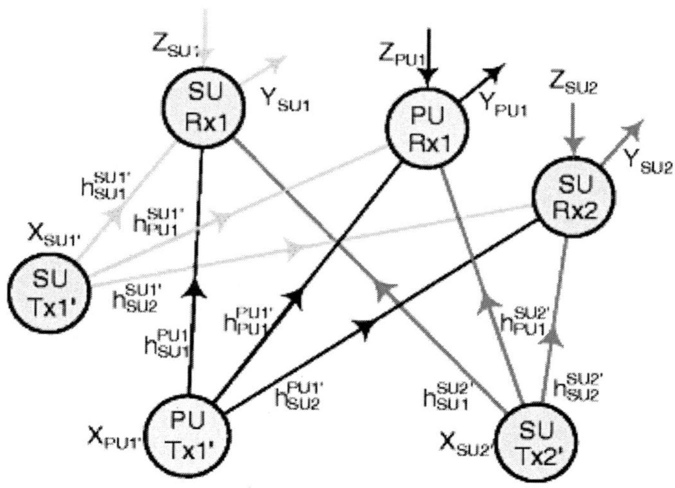

Figure 3 Channel model for scenario of co-existing PUs and SUs

The signal received at the PU or at the SU can be represented with a sum of: the desired received signal, interfering received signal, and the noise. Signal received at *i*th PU and signal received at *l*th SU is a superposition of various components and can be expressed as

$$Y_{\mathrm{PU}i} = h_{\mathrm{PU}i}^{\mathrm{PU}i'} X_{\mathrm{PU}i'} + \sum_{j'=1, j' \neq i}^{N_{\mathrm{PU}}} h_{\mathrm{PU}i}^{\mathrm{PU}j'} X_{\mathrm{PU}j'} + \sum_{k'=1}^{N_{\mathrm{SU}}} h_{\mathrm{PU}i}^{\mathrm{SU}k'} X_{\mathrm{SU}k'} + Z_{\mathrm{PU}i}, \tag{1}$$

$$Y_{\mathrm{SU}l} = h_{\mathrm{SU}l}^{\mathrm{SU}l'} X_{\mathrm{SU}l'} + \sum_{j'=1}^{N_{\mathrm{PU}}} h_{\mathrm{SU}l}^{\mathrm{PU}j'} X_{\mathrm{PU}j'} + \sum_{k'=1, k' \neq i}^{N_{\mathrm{SU}}} h_{\mathrm{SU}l}^{\mathrm{SU}k'} X_{\mathrm{SU}k'} + Z_{\mathrm{SU}l}, \tag{2}$$

where $X\mathrm{PU}i'$ is the signal from PU transmitter i', $X\mathrm{SU}l'$ is the signal from SU transmitter l', l', $h_{\mathrm{PU}i}^{\mathrm{PU}i'}$ denotes the channel coefficient between PU transmitter i' and PU receiver i, l', $h_{\mathrm{PU}i}^{\mathrm{PU}i'}$ is the channel coefficient between SU transmitter k' and PU receiver i, i, $h_{\mathrm{PU}i}^{\mathrm{SU}k'}$ is the channel coefficient between SU transmitter l' and SU receiver l, $h_{\mathrm{SU}l}^{\mathrm{PU}j'}$ is the channel coefficient between PU transmitter j' and SU receiver l, N_{PU} is the number of PUs, and N_{SU} is the number of SUs, as illustrated in the Fig. 3. $Z_{\mathrm{PU}i}$ and $Z_{\mathrm{SU}l}$ represent additive white Gaussian noise at the PU receiver i and SU receiver l with the variances $\sigma 2\mathrm{PU}i$ and $\sigma 2\mathrm{SU}l$.

The desired received signal strength represents the power of the useful signal received at the SU receiver from the wanted SU transmitter contributing to the information data flow. The interfering received signal strength is the superposition of all received interfering signals at the input of the same SU receiver. Accordingly, we can express the desired received signal strength $P_{\mathrm{Rx\_SU}l\_D}$, and interfering received signal strength $P_{\mathrm{Rx\_SU}l\_I}$ at SU receiver l as

$$P_{\mathrm{Rx\_SU}l\_D} = \left| h_{\mathrm{SU}l}^{\mathrm{SU}l'} \right|^2 P_{\mathrm{Tx\_SU}l'}, \tag{3}$$

$$P_{\mathrm{Rx\_SU}l\_I} = \sum_{k'=1,k'\neq l}^{N_{\mathrm{SU}}} \left| h_{\mathrm{SU}l}^{\mathrm{SU}k'} \right|^2 P_{\mathrm{Tx\_SU}k'} + \sigma_{\mathrm{SU}l}^2 + \sum_{m'=1}^{N_{\mathrm{PU}}} \left| h_{\mathrm{SU}l}^{\mathrm{PU}m'} \right|^2 P_{\mathrm{Tx\_PU}m'}, \tag{4}$$

where *P*Tx_PU*m*′, *P*Tx_SU*l*′, and *P*Tx_SU*k*′ are PU and SU transmitting powers, and $\left| h_{\mathrm{SU}l}^{\mathrm{SU}l'} \right|^2$ and $\left| h_{\mathrm{SU}l}^{\mathrm{PU}m'} \right|^2$ are the wanted and interfering channel gain coefficients. Received interfering power *P*Rx_SU*l_I*

takes into account the interference received from PUs, interference from other SUs on co-channel, and adjacent channel operating in the interference range of the corresponding SU and the noise. All those interferences add up to the cumulative interference, which reduces quality and reliability of the SU link. Similarly, we can calculate the desired received signal, and interfering received signal at the PU, which is omitted for brevity.

The SU throughput is proportional to the normalized ergodic channel capacity, which is defined as the maximum achievable rate averaged over all fading blocks and it can be expressed as

$$C\,(\mathrm{SINR}) = \max\, E\left[log_2 \left(1 + F^{-1}\,(1-\epsilon)\,\mathrm{SINR} \right) \right], \tag{5}$$

where $F(x) = P\left(\left| h_{\mathrm{SU}l}^{\mathrm{SU}k'} \right|^2 > x \right)$ is the complementary cumulative distribution function of SU channel gain $\left| h_{\mathrm{SU}l}^{\mathrm{SU}k'} \right|^2$, ε is an error probability of the secondary transmissions, and SINR is the desired signal-to-interference-plus-noise ratio at the SU receiver. If the SU channel gain is below minimal threshold level *x*, no useful data transmission can be obtained between the SU transmitter and receiver. In Eq. (5), the theoretical relation between SU throughput and an ergodic channel capacity for the fading channel is shown. In the numerical simulation in Section 8, we assume that SU system utilizes M-ary quadrature amplitude modulation (M-QAM) and we use the approximation (19) for SU throughput. Additionally, in the simulation, we have set up the upper limit of the achievable individual SU throughput in order to avoid unfair spectrum usage between the SUs.

4.2 Weighting coefficients characterization

Since all of the channel parameters and fading statistics between interfering nodes are generally unknown to the SUs, it would be demanding to implement full network interference model. On the other hand, reliable and precise interference estimation is important for constructing the CR conflict graph needed for the FA algorithm implementation. Therefore, to encompass the sensitivity to the potential levels of the interference in the process of assignment of the frequencies to the SUs, we propose the classification of the interference in four categories. This enables an efficient practical implementation of the FA protocol with control and quantification of the interference levels, while not adding too much of an additional burden to the network communication overhead. Considering the spectrum sharing between the SUs, for each interfering path between SUs we designate co-channel and adjacent channel conflict edge and introduce associated interference weights determining the level of the potential interference between the corresponding SUs as follows:

$$w_{\mathrm{co}} = p_{\mathrm{co}} \cdot w_i \quad , \quad w_{\mathrm{adj}} = p_{\mathrm{adj}} \cdot w_i, \tag{6}$$

where w_{co} and w_{adj} are co- and adjacent channel interference edge weights, p_{co} and p_{adj} are co- and adjacent channel penalties, and w_i is the nominal edge weight.

Co- and adjacent channel interference edge weights are the network penalties for violation of the soft FA constraints in the neighboring vertex, resulting in the deterioration of the objective function used for the optimization of the FA process. The channel penalties p_{co} and p_{adj} represent the portion of the CR transmitter power which is absorbed by the CR receiver on the same or adjacent channel. Generally, channel penalty is calculated as the reduction of the interference power caused by the filter shape of the transmitter spectrum density mask and the receiver selectivity mask or it can be measured. As we are analyzing network with a predefined set of frequencies and aligned channel bandwidths, in network simulation in Section 8, we assume that the channel penalties are predefined and have the constant values. The nominal edge weight w_i corresponds to the channel coefficient including propagation loss, antenna gain, and discrimination between the SU Tx and SU Rx under the consideration.

We propose to classify level of the potential interference in four main categories (*harmful*—high interference; *disturbing*—medium-high interference; *annoying*—medium-low interference; *permissible*—low interference) and assign the corresponding conflict edge weight coefficients $w_i \in \{0;0.4;0.7;1\}$. To classify the potential interference in the appropriate category, a value of the potential interference between the SU Tx and Rx is compared with the cumulative probability density function of the average received interference power between the SUs. The cumulative probability density function (cdf) is empirically determined from the previous spectrum sensing measurements in the frequency band under consideration. On the basis of this comparison, it is identified to which category out of four this potential interference belongs and the characteristic value of the nominal edge weight factor is assigned for the identified interference category. Proposed values for the categorization of the nominal edge weights are calculated by evaluating the cdf of the average received interference power between the SUs. Interference cdf is divided into four non-overlapping areas and a median cumulative probability value for each area is calculated. Channel weights representing each category are calculated as the ceiling round up (rounded off to the higher value) for the edge weight coefficients (0.4;0.7;1) and as floor round up for edge weight coefficient (0). Specific weights for four different categories of the interference are chosen in such way that the possible harmful interference between SUs is emphasized. We have tested proposed algorithms with the variation of the values of discrete interference weights, and the results show similar behavior as the algorithms with a small variation of the nominal network weight coefficients representing four categories.

This novel interference categorization allows the introduction of the susceptibility of the FA protocol to the level of the interference, taking into account the probabilistic nature of the received signal as described above, as well as possible interference estimation errors.

To illustrate the used model of the interference classification, Fig. 4 shows the criteria for the forming of a list of blocked channels B_v, depending on a measured level of the PU signal at the corresponding SU. I_{PU} is the level of the interference from PUs transmissions at the observed frequency measured on the SU terminal. If the value of I_{PU} is above the threshold of the receiver interference limit, the said frequency cannot be temporarily used for the SU transmissions, since it may cause harmful interference to the PUs in the vicinity of the SU terminal. The frequency is included in the dynamic list of the blocked channels at this SU. If I_{PU} is below the threshold, the interference is small and SU can share the frequency with the PUs with appropriate transmission power. Probability density function of the received interfering signal is shown in the middle of Fig. 4. In the right side of Fig. 4, a proposed categorization of the SU interference corresponding to SU interference level, edge weights coefficients, and their binary coding is shown. I_{SU} is the level of the measured interference from the SU transmissions, used for determining the SU adjacency and the conflict edge weights. The SUs edge weights are used for quantitative estimation of the interference in the process of the assigning frequencies. In such a way, we can make more knowledgeable frequency decisions and control individual interference components as well as aggregate interference from multiple users during the FA process. In the proposed FA algorithms, we use the continuous edge weights in the interval [0,1], or the discrete weight values as shown in Fig. 4. The protocol for determining the SUs adjacency and the conflict graph edge weights is elaborated in Section 6.

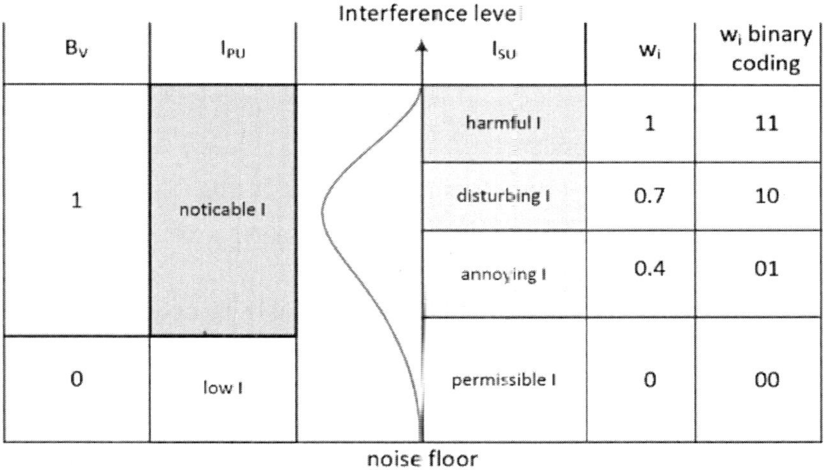

Figure 4 Categorization of interference for FA in CR networks. The criteria for forming list of blocked channels is shown on the *left* side. Classification and quantification of SU interference is shown on the *right* side

5 INTERFERENCE SENSITIVE FA ALGORITHMS

The proposed FA in the CR networks is based on the efficient heuristic based algorithms that can run reasonably fast and provide a good quality solution. In this paper, non-uniform traffic on the different communication links and frequencies, weighted interference model, and non-orthogonal channels are considered based on the CR networks specifics.

5.1 Algorithm description

In a proposed model, we have a predefined set of frequencies F={1,..,M} and the task is to assign frequencies to all SUs requesting network access, taking into consideration the network interference. The SU is not allowed to use any of the frequencies which could cause noticeable interference to the PU network, determined by the spectrum awareness and expressed with a set of blocked channels B_V. If there are no available channels in the spectrum pool, SU's request is rejected or transferred to another spectrum band. In our approach, the SUs connected with the conflict graph edge (i.e., interfering each other) can share the same frequency or an adjacent frequency, but with a corresponding penalty, which is proportional to a level of the mutual interference calculated as a product of the nominal edge weights and co-/adjacent channel penalty. Having this quantitative metric, resource manager selects the frequencies with a minimal interference contribution towards the other users. Proposed FA algorithms consist of four main phases with specific tasks in each phase as follows:

- Phase 1. FA preparation:
    - Establish the list of the blocked channels at all SUs.
    - Determine the SU transmitting power for each available channel at all SUs.
    - Calculate the SU throughput for each available channel at all SUs.
    - Establish a conflict graph edges and determine the interference weights for co-channel and adjacent channel transmissions.
    - Interference categorization.
- Phase 2. Selecting the next SU to assign frequency:
    - Label the non-assigned SUs using specific CR FA saturation metric.
    - Dynamically order the SUs with decreasing saturation score.
    - Select the SU with the highest saturation metric to assign frequency as the next SU to process.
- Phase 3. Assigning frequency to the selected SU:
    - Calculate objective function for each available frequency at the selected SU.
    - Select and assign frequency maximizing objective function at the selected SU.
- Phase 4. CR FA performance evaluation and conflict graph update:
    - Calculate interference, throughput, and throughput variance for all SUs.
    - Determine the overall network performance and update conflict graph.

The initial SU transmitting power for each available channel at SUs is determined in a way that SU does not cause harmful interference to the closest PU using this channel and that the minimal required SINR at the SU user is achieved. In the following phases, the SU calculated transmitting power per channel is updated using power control ratio parameter. The details of the power control mechanism are described in Section 5.3, while establishing the list of blocked channels, calculation of the conflict graph edge weights, and its categorization are performed as described in Section 4.2.

We propose the centralized and distributed algorithms with the objective of minimizing the cumulative CR network interference and with the objective of maximizing the network throughput, as shown in Table 3.

Table 3 Summary of proposed FA algorithms

	Objective	
Implementation	Minimal interference	Maximal throughput
Centralized	CminSumIrt	CMaxSumCap
Distributed	DminInt	DMaxCap

5.2 Saturation metric labeling and coloring

The proposed algorithms process and assign frequencies to the SUs requiring spectrum access in a sequence where in each iteration step the most difficult SU is selected. Most difficult SU is the vertex with the smallest choice of channels determined by the saturation metric. The saturation metrics label S_{vi} is calculated as

$$S_{\text{vi}} = \sum_{\text{vi} \in V_c} B_{\text{vi}} + \sum_{f \in F,\ \text{vj} \in V_c} E_{\text{co vi vj}} \cdot x_{\text{vj}} \cdot w^{\text{co}}_{\text{vi vj}} + \sum_{f \in F,\ \text{vk} \in V_c} E_{\text{adj vi vk}} \cdot x_{\text{vk}} \cdot w^{\text{adj}}_{\text{vi vk}}, \tag{7}$$

where B_{vi} are blocked frequencies, which cannot be used at the SU vi, $E_{\text{co vi vj}}$, and $E_{\text{adj vi vk}}$ are co- and adjacent channel edges indicating the interference between the SUs, *wvi* vjco and *wvi* vkadj are the conflict edge weights determining the interference potential of the adjacent SUs, x_{vj} and x_{vk} are variables indicating that the adjacent SUs have the frequency assigned. All the vertices are sorted in the decreasing order of the saturation label. The vertex with the largest label is selected for coloring. Saturation metrics score is changing in each iteration step due to the continuous process of the assigning frequencies to the SUs, and therefore, the order of the SUs also dynamically changes. In the assignment phase, the selected SU is colored with a color which contributes the most to the system utility function. Here, we investigate the different algorithms with the objective of the minimal interference and maximal throughput.

5.2.1 Minimum cumulative network interference FA

The proposed algorithm is minimizing the total network interference while taking into account co-channel and adjacent channel interference. Vertex coloring is performed according to

$$\mathrm{Col}_{\mathrm{vi}} = \operatorname{argmin}\left[\alpha + \beta\right], \tag{8}$$

$$\alpha = \sum_{vj \in E_{\mathrm{co\,vi\,vj}}} w^{co}_{\mathrm{vi\,vj}} \cdot x_{\mathrm{vj}}, \tag{9}$$

$$\beta = \sum_{\mathrm{vk} \in E_{\mathrm{adj\,vi\,vk}}} w^{\mathrm{adj}}_{\mathrm{vi\,vk}} \cdot x_{\mathrm{vk}}, \tag{10}$$

where α and β are co- and adjacent channel cumulative interference at the adjacent vertices. In coloring, the algorithm is selecting the frequency which contributes the least to the cumulative network interference. Having the interference represented by edge weights, the individual interference contributions can be quantified and the level of total network interference can be controlled.

5.2.2 Interference sensitive maximum throughput FA

The FA is performed using a comprehensive coloring argument that integrates the interference and throughput. The proposed algorithm is maximizing the total network throughput while taking into account co-channel and adjacent channel interference. The coloring argument is calculated as

$$\mathrm{Col}_{\mathrm{vi}} = \mathrm{argmax}\left[\frac{\sum C_{f\,\mathrm{vi}} \cdot x_{\mathrm{vi}}}{\sum_{\mathrm{vi} \in V_c} B_{\mathrm{vi}} + \alpha + \beta}\right], \tag{11}$$

where $C_{f\,\mathrm{vi}}$ is the SU throughput at frequency f. The SU throughput is calculated using Eq. (5), and its value can be between 0 and the maximum available throughput per SU. This value is set to 16 Mbit/s in the numerical simulation in Section 8. The coloring is done with the preference given to the frequencies having the highest throughput, while having the minimal influence to the rest of the SU network.

5.3 Transmit power control

The proposed transmit power control strategy is based on the balancing of the SU transmit power level between minimal required to satisfy QoS requirements and maximal permissible level in order to avoid interference with the PUs. In order to minimize the interference to the other concurrent SUs, it is recommended to maintain the transmission power level at a minimum, while ensuring an adequate signal quality at the receiving end. Minimum required transmit power is determined by adjusting the power level until the targeted SINR at the SU receiver is satisfied. Maximum acceptable transmit power is established by considering the permissible interference at the PUs.

The initial SU power is calculated using the path loss estimation determined by spectrum sensing of the PUs as:

$$P_{\mathrm{Tx\_SU}} = P_{\mathrm{Tx\_C}} - P_{\mathrm{Rx\_C}} + S_{\mathrm{R}} - L_{\mathrm{A}} + P_{\mathrm{Rx\_SU\_I}}, \tag{12}$$

where $P_{\mathrm{Tx\_SU}}$ is the initial SU transmit power, $P_{\mathrm{Tx\_C}}$ is the predetermined transmitted power of the common pilot channel, $P_{\mathrm{Rx\_C}}$ is the measured power of the common pilot channel at the SU receiver, L_{A} is the additional gain, loss, and tolerances, S_{R} is the required SINR, and $P_{\mathrm{Rx\_SU\_I}}$ is the interfering power. The initial power level is used for the initial communication between SU terminals and as a starting power level for the process of adaptive adjusting of the transmitter power. Instant transmit power of the secondary user is determined as

$$P_{\mathrm{Tx\_SU}} = \left\{\begin{array}{ll} P_{\mathrm{Tx\_SU\_R}} & \text{if } P_{\mathrm{Tx\_SU\_M}} \geq P_{\mathrm{Tx\_SU\_R}} \\ 0 & \text{if } P_{\mathrm{Tx\_SU\_M}} < P_{\mathrm{Tx\_SU\_R}} \end{array}\right\}, \tag{13}$$

where $P_{\mathrm{Tx\_SU\_M}}$ is the maximal acceptable transmit power and $P_{\mathrm{Tx\_SU\_R}}$ is the required transmit power. The maximal acceptable transmit power

Algorithm 1 :
Input:
Conflict graph $G_c(V_c, E_{co}, E_{adj})$, set of frequencies F, set of blocked frequencies B_v, set of co-channel and adjacent channel edge weights w_{co}, w_{adj}, penalties for co- and adjacent channel rule violation p^{co}, p^{adj}, set of transceiver user throughputs c_v
Output:
Frequency assignment function $f_{best}: V_c \to F$, cumulative network interference I_{sum}, total SU network throughput C_{CRN}
begin
initialize all
***do while** not all vertices V_c have been colored*
    ***for** $\forall V_c$ calculate saturation metrics label S_{vi}*
    ***sort** vertices in decreasing order of S_{vi}*
    ***select** vertex with Max S_{vi}*
    ***color** vertex vi according to objective function*
    update all
repeat
$f_{best} = x_{vf}$
***calculate** I_{sum}, C_{CRN}*
end

Figure 5 Centralized FA algorithm. The figure shows pseudo code of proposed centralized algorithm

represents the maximal allowed power of the cognitive SU in order to satisfy the interference constraints, as illustrated in Fig. 4, and is determined by the spectrum sensing of the PU at the SU location. The required transmit power is the transmit power level just strong enough to satisfy SINR for obtainable throughput at the SU receiver. The comparison of the measured SINR at the SU receiver with the required SINR determines the SU transmit power control ratio for the transmit power adjustment. The new minimum required transmit power is obtained by multiplying present minimum required transmit power with the transmit power control ratio R_{TPC} as follows:

$$P_{\mathrm{Tx\_SU\_R}}(t+1) = P_{\mathrm{Tx\_SU\_R}}(t)\, R_{\mathrm{TPC}}. \tag{14}$$

A possible implementation of the proposed adaptive transmit power control strategy using fuzzy logic power controller is described in [23].

6 CENTRALIZED AND DISTRIBUTED ALGORITHM

6.1 Centralized FA algorithm

The centralized network resource manager has the advantage of collecting all the relevant information and using more complex processing engines. It can, therefore, provide an overall better solution. The centralized algorithm can keep track of the CR user mobility, achieving continuous connectivity of the established CR session. Pseudocode of the centralized FA algorithm is presented in Fig. 5.

In each iteration, the centralized algorithm calculates the saturation metrics of all unassigned vertices. In order to detect the most difficult vertex to process first, the vertices are sorted and vertex with the largest saturation metrics is selected. For the selected SU, coloring is performed with the frequency which maximizes the objective function. In CminSumInt algorithm, the frequency is selected on the basis of the minimal interference caused to the adjacent vertices. In CMaxSumCap algorithm, the selected frequency maximizes the comprehensive metric that integrates both interference and throughput. The centralized solution requires an additional communication overhead to collect and distribute information between the vertices. In that respect, the common communication channel and protocol are necessary. Also, the centralized solution is sensitive to the potential failures of the central entity, which can result in the discontinuities of the FA in the CR networks. In the cluster topology, the CR network is subdivided in a number of geographically distinguished clusters. Each cluster uses a different resource manager and an independent algorithm determining the next SU to assign frequency using the saturation metric label and frequency assignment maximizing the network objective.

In order to determine the complexity of the algorithm, it is essential to determine the number of iterations needed for the algorithm to terminate. The number of iterations in the centralized algorithm is bounded by $O(N_{\mathrm{SU}})$, where N_{SU} is the number of the secondary users requesting access to the CR network. Since the single vertex coloring can be completed in $O(M)$ where M is the number of available frequencies, the overall complexity of the proposed centralized algorithm is $O(N_{\mathrm{SU}} M)$.

6.2 Distributed FA algorithm

The distributed algorithm finds a local objective, but it results in a more robust, flexible, faster and less communication demanding implementation. Proposed distributed algorithm works on a vertex level and each vertex only has the local information, which is communicated between the adjacent incident vertices. Pseudocode of the distributed FA algorithm is shown in Fig. 6.

Algorithm 2 :
Input:
Set of blocked frequencies, vertex co-channel and adjacent channel edge weights w_{co}, w_{adj}, penalties for co- and adjacent channel rule violation p^{co}, p^{adj}, vertex transceiver user throughputs c_v
Output:
Frequency assignment function $f_{best}: V_c \rightarrow F$ for vertices incident on v_i, cumulative vertex interference I_{sum}, SU throughput C_{CRN}
begin
initialize all
***calculate** saturation metrics label S_{vi} for vertex v_i*
***calculate** vertex contention window for backoff time*
***transmit** saturation metrics label S_{vi}, blocked frequencies vector B_{vi}*
***receive** S_{vi}, B_{vi} of all incident vertices*
***do while** not all V_c incident to vertex v_i are colored*
    ***sort** vertices in decreasing order of S_{vi}*
    ***select** vertex with Max S_{vi}*
    ***if** vertex v_i has Max S_{vi}*
      ***select frequency** with maximum objective*
      ***color** vertex vi according to objective function*
    endif
    update all
repeat
$f_{best} = x_{vf}$
***calculate** I_{sum}, C_{CRN} for v_i*
end

Figure 6 Distributed FA algorithm. The figure shows pseudo code of proposed distributed algorithm

The algorithm initializes and the SU requesting frequency access calculates its own saturation metrics label. Vertex saturation label and set of blocked frequencies are communicated to other adjacent non-assigned vertices requesting access to the same frequency band. In order to avoid transmission collision, contention window for the transmission backoff time is calculated as a random value in the interval [0,window], where the window is calculated as $1/S_{vi}$. On the basis of the exchanged information on the local vertex saturation metric, ordering of the vertex coloring is determined. Vertex with a local maximum saturation score is selected first to color, and then it selects a frequency with the maximum contribution to the objective function. Selected frequency is then communicated to the adjacent vertices and the list of assigned frequencies is updated. Subsequently, new saturation labeling is performed and the next vertex to color is determined.

Each iteration of the single vertex coloring can be completed in $O(M)$ where M is the number of the available frequencies. The number of the iterations in the distributed algorithm is bounded by $O(\Delta)$, where Δ is a maximum degree of a vertex in the conflict graph. The overall complexity of the proposed distributed algorithm is $O(\Delta M)$. The complexity of the distributed algorithm is N_{SU}/Δ times smaller compared to the centralized algorithm, indicating that the distributed algorithm is simpler and converges faster.

7 GENERALIZATIONS AND IMPLEMENTATION

7.1 Central frequency and bandwidth CR spectrum assignment

In the proposed FA, we assume fixed set of the channels with predetermined bandwidth and channel separation. In general, the problem of the FA in the CR networks should include not only the identification of the central frequency but also the optimal bandwidth to be used according to the service requirements of the SUs. Generally, spectrum holes can be of the various bandwidth and frequency separation, and CR devices by definition should be able to use these spectrum opportunities. The question of the assigning spectrum fragments of the different sizes is still an open area in research of the CR FA [10]. In this section, we present the generalization of the proposed model of the FA in order to relax input assumptions of the fixed channel division, to an implementation considering bandwidth and central frequency decision. This generalization is quite useful in the practical deployments of the CR systems.

The proposed solution of the central frequency and bandwidth assignment for the CR consists of the following steps: the admission control, the bandwidth reservation, and the spectrum assignment. In the admission control, the spectrum management system has to determine weather the new incoming CR communication request can be accepted or the request is queued. Fair spectrum bandwidth B_{SUi}, which could be assigned to the SUi is calculated as

$$B_{\mathrm{SU}i} = \frac{C_{\mathrm{SU}r\_i} B_{\mathrm{free}}}{\sum_{j=1}^{N_{\mathrm{SU}}} C_{\mathrm{SU}r\_j}}, \tag{15}$$

where CSUr_iand CSUr_jare the requested throughputs of SU i and SU j and B_{free} is the available idle spectrum in the frequency band considered and not currently used by PUs.

The SUi request is accepted, if the achievable throughput in the bandwidth which can be assigned to the SU can support the minimal sustainable throughput needed for satisfying the service requirements for the SU as:

$$\mathrm{ack}_{\mathrm{SU}i} = \left\{ \begin{matrix} 1 \ if \ C_{\mathrm{SU}a} = B_{\mathrm{SU}i} \cdot c_{\mathrm{SINR}} \geq C_{\mathrm{SU}s\_i} \\ 0 \ if \ C_{\mathrm{SU}a} = B_{\mathrm{SU}i} \cdot c_{\mathrm{SINR}} < C_{\mathrm{SU}s\_i} \end{matrix} \right\}, \tag{16}$$

where ack_{SUi} is the binary acceptance variable, C_{SUa} is the achievable SU throughput, c_{SINR} is the normalized channel throughput of the SU with observed SINR and C_{SUs_i} is the minimal sustainable throughput. Achievable SU throughput of the SU is calculated using Eq. (5) or by approximation Eq. (19) in case of M-QAM.

For the bandwidth reservation, we propose to use MSA algorithm of the spectrum aggregation presented in [24]. The MSA algorithm is utilizing the worst spectrum band which can barely satisfy the bandwidth requirements B_{SUi} of the SU in consideration relating to all unassigned SUs. In MSA algorithm, the SUs are sorted by the bandwidth requirements and the idle spectrum is sorted by the available bandwidth of the spectrum regions. For the bandwidth reservation, the MSA algorithm firstly chooses the SU with the highest bandwidth requirements and try to assign the spectrum region with the least available bandwidth. The reservation continues in the decreasing order of SU's bandwidth requirements and ascending order of the available spectrum regions, until all users are assigned or there is no available spectrum.

The spectrum assignment phase can use proposed algorithms for the CR FA presented in this paper without the need for modification. In the preparatory phase, generalization of the interference weighting to include interference potential between the spectrum fragments is proposed. As we assign discontinuous spectrum regions of the different bandwidths to SUs, we have to extend the definition of the spectrum weights to be able to accommodate the evaluation of the level of the interference between the SUs. For each interfering path between the SUs, we designate the conflict edge and introduce the associated interference weight w_E determining the level of potential interference between corresponding SUs as follows:

$$w_E = p_E \cdot w_{N0}, \tag{17}$$

where p_E is the conflict edge channel penalty and w_{N0} is the nominal edge weight. The nominal edge weight w_{N0} corresponds to the channel coefficient including propagation loss, antenna gains, and discrimination between the SU Tx and SU Rx under the consideration. It is determined as described in Section 4. The conflict edge channel penalty p_E represents the portion of the SU transmitter power which is absorbed by the SU receiver. It is calculated as the reduction of the interference power caused by the filter shape of the transmitter spectrum density mask and the receiver selectivity mask as follows:

$$p_E = \int_{f_l}^{f_h} S_{SUTx}(f) \cdot S_{SURx}(f - \Delta f) \cdot df, \tag{18}$$

where S_{SUTx} is the Tx signal's power distribution across the frequency spectrum, S_{SURx} is the SU receiver filter frequency response, f is the frequency, f_l and f_h are lower and higher limits of the frequency band in consideration, and $\Delta f=|f_{Tx0}-f_{Rx0}|$ is the spectrum gap between the SU Tx and Rx center frequency. If $\Delta f=0$, both Tx and Rx are operating at the same central frequency and we are coming to the special case of co-channel penalty.

7.2 Implementation considerations

The CR networks utilizing algorithms with the objective of the minimal total network interference (CminSumInt, DminInt) are more suitable for the user scenarios where there exists a large number of SUs and scarcity of the available frequencies, requiring reliable but not very data demanding communications. The real-world applications of CR networks with this objective function could be the wireless sensor networks, the radiometers in the smart grid networks, the medical body area network devices for monitoring, diagnosing or treating of the patients, the utility companies' telemetry networks, location services in the search and rescue applications, different machine to machine communications, backup system for the emergency networks, and the military communications in the hostile interference environment. The CR networks using algorithms with the objective of the maximal throughput with the interference susceptibility (CMaxSumCap, DMaxCap) are more appropriate for the capacity oriented applications encompassing different scenarios including internet connectivity, cellular networks, and TV white space applications. The real-world applications could be the cellular networks data offload, realizing femtocells in the cellular networks, the best effort wireless hotspots, the local internet connectivity, the internet of things applications, the public protection and disaster relief networks realizing communication between the emergency respondents and the public safety agencies, and the wireless cameras video transfer.

The FA in the CR is an event-based functionality and continuously re-iterated process. In a given moment in time, frequencies for only part of the active SUs have to be assigned. When the PU appears, all active SUs in the interference range have to move to the new spectrum bands [11]. This event triggers re-initiation of the FA process for multiple SUs. When new SU appears in the CR network, it needs to be assigned frequency for its transmissions. Similarly, when the quality of the SU transmissions deteriorates due to dynamic channel conditions, the SU wants to switch to a better frequency. In those cases, the process of a single frequency selection is initiated.

The network protocol for determining the conflict graph edges and corresponding weights can be as follows. Each of the active SUs interchangeably transmits SU ID and test message on the common communication channel, while all other SUs are temporarily monitoring the channel and receiving. The test message and default transmission parameters of the SU are known to all of the SUs in the same geographic area. By monitoring the received signal strength, each of the SUs can determine its adjacent SU transceivers and categorize the interference potential with the corresponding interference category (harmful, disturbing, annoying, permissible). Accordingly, each SU can construct a local conflict network graph with edges corresponding to all of the received SUs transmissions and associated weighting coefficient proportionate to the received signal strength. In the centralized implementation, these network weights are then communicated to the central spectrum management entity, which constructs the network graph and assigns frequencies. The implementation of the protocol determining network conflict graph and corresponding weights is easy to implement and it can be coded by only 2 bits. Since determining of the conflict graph edges and corresponding weights adds additional communication overhead to the CR communication, it is rational to construct the conflict graph and perform weight determining protocol only when needed. In the case of the stationary SU, the weight determining protocol should be performed at the time of the establishing the network or when some change in the network configuration or propagation parameters occurs, while in the case of moving SUs, the weight determining protocol should be performed more often. This could be determined by a network policy and triggered when the CR terminals determine the increased probability level of dropped or severely disturbed transmissions.

8 NUMERICAL RESULTS AND DISCUSSION

8.1 Simulation setup

In order to evaluate the performance of the proposed FA algorithms, we have implemented the stochastic simulation of the primary and the CR network. In the setup of the stochastic simulation, we assumed that the area under test is a square with dimensions of 30 km × 30 km. We considered the SU Tx –Rx pairs that co-exist with a cellular PU network in a configuration similar to the one illustrated in Fig. 2. PU network consists of a number of base stations covering a circular area around them. CR network consists of N_{SU} short range secondary links which access radio spectrum opportunistically in the same area. The frequency band considered is 2000 MHz, and available frequencies are consisting of the group of 15 to 25 channels with a channel bandwidth of 3.5 MHz. The number of the PUs active in the area under test is between 15 and 40; the number of the SU links requesting frequency access is between 10 and 50. The whole spectrum is considered to be open for an opportunistic spectrum access scenario, meaning the SUs can use any channel on the condition of not causing harmful interference to the PU network. We summarize the simulation parameters in Table 4.

PU network model. In the simulation, we considered the network of the PU base stations, each of them operating on one channel, which has to be protected by all SUs in the corresponding interference range. We have randomly placed a number of PUs in a given area. For simplicity, we assumed that the PUs have a constant Tx power of 43 dBm and omnidirectional antenna radiation pattern. Each one of the PUs is defined as an information source, which can be in two distinctive states: an active state in which PU constantly generates information packets and requests for data transmission or in a passive state in which PU is not active as an information source. The PU entity is modeled as a two state Markov birth-death process with birth rate α and death rate β. Since each user arrival is independent, transition follows the Poisson arrival process.

Table 4 Simulation parameters

Parameter	Value
Area under test	30 km × 30 km
Number of PUs (N_{PU})	15–40
Number of SUs (N_{SU})	10–50
Frequency band	2000 MHz
Channel bandwidth	3.5 MHz
Number of frequencies (M)	15–25
PU transmission range (dPU)	10 km
SU transmission range (dSU)	1–4 km
PU interference range	20 km
SU interference range	2–8 km
Modulation M-QAM	M = [4, 16, 32, 64]
Maximal SU throughput (64 QAM)	16 Mbit/s

CR network model. Cognitive secondary system co-exists in the same area with a primary system using opportunistic radio spectrum access. The SU transmitter and receiver pairs are uniformly randomly generated in the analyzed area. The distance between SU Tx and SU Rx is set between 1 and 4 km. In order to test proposed algorithms in a difficult sharing environment with many interference cases and for simplicity reasons, we assume that SUs also have an omnidirectional antenna radiation pattern. The SU adjusts its transmitting power in order to avoid generating interference with the active PUs on the same frequency. The SU Tx power is limited to a maximal value of P_{SUMAX}=27 dBm, and it can have any continuous value in the interval [0,P_{SUMAX}]. The SU Tx power is calculated for each available channel using the transmit power strategy as described in Section 5.3. For the channels which are included in the local list of blocked channels B_v, the SU TX power is set to 0 because they cannot temporarily be used. In the simulation, we assumed that the SUs use square M-QAM with Grey bit mapping, without losing the generality of the proposed algorithms. For the calculation of the SU throughput in simulation, we used the approximation in [25], tight within 1 dB for M≥4 and BER≤10^{-3} as

$$C(\text{SINR}) \approx \text{BW}_{\text{ch}} log_2 \left[1 + \frac{\text{SINR}}{-ln(5\text{BER})/1.6} \right], \tag{19}$$

where $C(\text{SINR})$ is the throughput, BW_{ch} is the channel bandwidth, SINR is the wanted signal to interference plus noise ratio at the SU receiver and $\text{BER}=10^{-3}$ is the target bit error rate. The SINR at each SU differs and depends on the channel used, obtainable SU Tx power, distance to SU Rx, and sum of all interference from other SUs. The maximal throughput of the individual SU is set to 16 Mbit/s, using 64-QAM modulation, as an upper limit of the achievable individual SU throughput. For each SU, the list of blocked channels is established, comparing calculated distance between the PU and SU with the PU interference range. If the distance is smaller than the interference range, this channel cannot be used at the SU. The corresponding channel is included in the SU list of blocked channels B_{v} in order to protect PU transmissions. This list for each SU is dynamically updated with each iteration of the FA algorithm. In the simulation, the SU edges and their corresponding co-channel and adjacent channel weights are calculated from the level of overlapping of interference ranges of the SUs as follows:

$$w_{\text{co}} = p_{\text{co}} \cdot w_i \quad , \quad w_{\text{adj}} = p_{\text{adj}} \cdot w_i, \tag{20}$$

$$w_i = \left\{ \begin{array}{cc} 1 & if\ d_{\text{SU}j}^{\text{SU}i} < 2d_{\text{ST}} \\ \frac{2d_{\text{ST}} - d_{\text{SU}j}^{\text{SU}i}}{2\left(d_{\text{SI}} - d_{\text{ST}}\right)} + 1 & if\ 2d_{\text{ST}} \leq d_{\text{SU}j}^{\text{SU}i} \leq 2d_{\text{SI}} \\ 0 & if\ 2d_{\text{SI}} < d_{\text{SU}j}^{\text{SU}i} \end{array} \right\} \tag{21}$$

where $d_{\text{SU}j}^{\text{SU}i}$ is the distance between SUi and SUj, d_{ST} is the SU transmission range, d_{SI} is the SU interference range, p_{co} is the co-channel penalty set to 1, p_{adj} is the adjacent channel penalty set to 0.1. Interference is classified in four categories as described in Section 4.2.

Methodology. The three parameters, the number of the available frequencies M, the number of the secondary users N_{SU}, and the number of the active primary users N_{PU} are tunable in the ranges shown in Table 4. Each time, we change one of the parameters in the tuning range and then compare the algorithms using the following metrics:

- Average throughput per SU (Mbit/s): the ratio of the cumulative CR network throughput over the number of active SUs
- Average interference per SU: the ratio of the cumulative interference of all interference sources over the number of active SUs
- SU throughput fairness: Jain's fairness index as a measure of the throughput distribution of all of the active SUs in the CR network

Instead of using the cumulative network values, we use the average throughput and average interference as a metric, to be able to compare the results of the simulations with different input parameters. The higher the average throughput per SU, the better the result. A smaller average interference indicates that the algorithm is better in respect of protecting the other users and more efficient in the spectrum usage. The fairness of the proposed algorithms is measured using Jain's fairness index (JFI) [26] calculated as

$$\mathrm{JFI}\left(x_1, x_2, \ldots, x_n\right) = \frac{\left(\sum_{i=1}^{n} x_i\right)^2}{n \sum_{i=1}^{n} x_i^2}, \tag{22}$$

where n is the number of the SU and x_i is the quantity of resources (i.e., throughput in our case) allocated to each user i. The value of JFI varies from $1/n$ representing poor fairness to 1 representing excellent fairness. Typically, systems with JFI larger than 0.5 are considered to provide a good fairness of the resource distribution.

On the basis of the simulated PU and CR network, we assign frequencies to each of the requesting SUs, using centralized and distributed FA algorithms under investigation. As a benchmark algorithm, we use the centralized and distributed version of Collaborative Max Sum Reward (CSUM) algorithm presented in [13], with a difference of assigning only one frequency per CR. As the performance of graph coloring depends on the conflict graph topology, we test algorithms on a large number of different networks for each setup of the input parameters, in order to have network neutral performance results. Each deployment of the PUs and SUs produces different network topology and a different corresponding conflict graph. All simulations are repeated 500 times with the selected set of input parameters. In each iteration, the new set of PUs and CR Tx-Rx pairs are generated, frequencies are assigned to the PUs, transmit power and throughput are calculated for all SUs, communication and conflict graph is constructed with corresponding interference weighting and categorization.

After that, the process of assigning frequencies to all SUs is performed with the proposed algorithms and the experimental performance of the algorithms is compared to the benchmark algorithms. The results are averaged to reduce the influence of the network topology or simulations variance on the network performance and presented results.

8.2 Overall FA algorithms results

In this section, we analyze and present the performance of the FA algorithms in a simulated environment of the PU and CR network. Figure 7 shows the cumulative distribution function $F(x)$ of the average interference level and average throughput per individual SU for 2000 iterations of the FA process with the same input parameters.

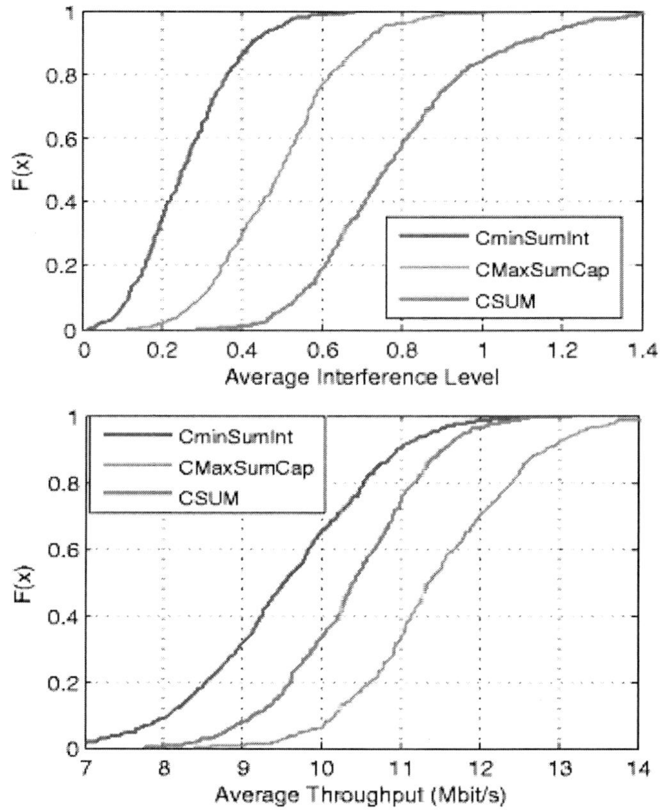

Figure 7 Cumulative distribution functions of average interference level and throughput (NoPU = 25, NoSU = 40, M = 15)

The CR network using centralized interference minimizing algorithm (CminSumInt) for assigning frequencies results in the lowest interference with the other concurrent users but also results in a lower individual throughput reward comparing to other algorithms. Throughput maximization algorithm (CMaxSumCap), on the other hand, outperforms all other analyzed algorithms taking into account the achieved SU throughput but causes more interference with the other users compared to a minimum interference algorithm. Figure 7 also shows tradeoff of the average interference level and the average throughput for the proposed algorithms. Using cumulative CR network interference minimization as the objective for FA also results a downturn in the lower level of average throughput compared to the algorithm using throughput maximization as the objective. For example, for the 90 % of the SUs in the network utilizing CminSumInt algorithm, the average interference level is below 0.4, while the average throughput is higher than 8 Mbit/s. On the other side, for the 90 % of the SUs in the network using CMaxCap algorithm, the average interference is below 0.7, and the average throughput is higher than 10.2 Mbit/s. The cumulative distribution functions in Fig. 7 show that labeling and coloring rules have been appropriately selected to achieve the minimal interference and maximal throughput objective. In the analyzed case, the network using CSUM algorithm have 2.5–3.5 times larger average level of interference per user compared to the network using CminsumInt algorithm. Introducing the interference sensibility in the proposed FA algorithms with co-channel and adjacent channel interference modeling, edge weights coefficients, and interference categorization significantly contribute to the algorithm efficiency related to interference reduction compared to benchmark CSUM algorithm, which employs binary interference model.

Figure 8 shows the convergence time required by the centralized and distributed algorithms to perform the FA, depending on the number of the SUs, for a network with 25 active PUs and 15 frequencies. As the number of SUs accessing the FA system increases, the time required to assign the frequencies to all SUs also increases. The convergence time for the centralized algorithm grows much faster than for the distributed algorithm and also shows a non-linear increase related to the number of SUs. On the other side, the distributed algorithm converges much faster than the centralized algorithm and shows a linear dependence on the number of SUs. The centralized algorithm has the higher complexity, since more iterations are needed compared to the distributed algorithm. The distributed algorithm performs local optimization involving adjacent vertices, and the FA process is done in parallel by the several SUs. As a result, differences in convergence times between algorithms become more significant when the number of the SUs increases. For the network with 50 SUs, the distributed algorithm is two times faster than its centralized counterpart. As we can see from Figs. 11 and 12, proposed centralized algorithms outperform distributed algorithms, but the time for calculating the FA scheme for all SUs is much longer. As a result, for bigger networks having more than hundred of SUs, usage of the centralized algorithms would become impractical. Due to the longer convergence time, the SUs would be in a situation of missing some of the spectrum opportunities (e.g., short temporal spectrum holes). We can conclude that, for the smaller CR networks, or for longer lasting spectrum opportunities (e.g., TV white spaces), the centralized algorithm is more suitable. On the other side, distributed algorithms would be a better choice for the bigger SU networks and spectrum band with more dynamic changes in the frequency availability.

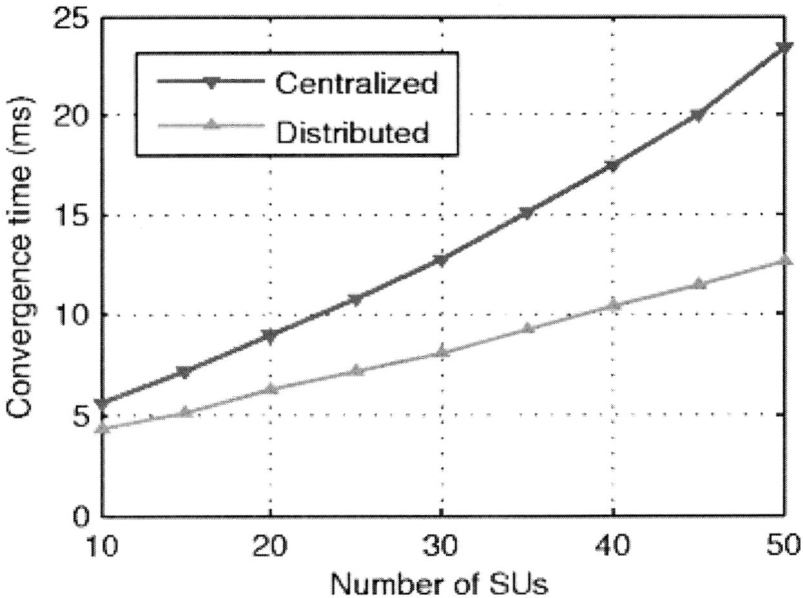

Figure 8 Convergence time for centralized and distributed algorithm. The figure shows performance metrics of proposed centralized and distributed algorithm

8.3 Centralized FA algorithms performance

The performance of the centralized FA algorithms in the CR networks is presented in Figs. 9 and 10.

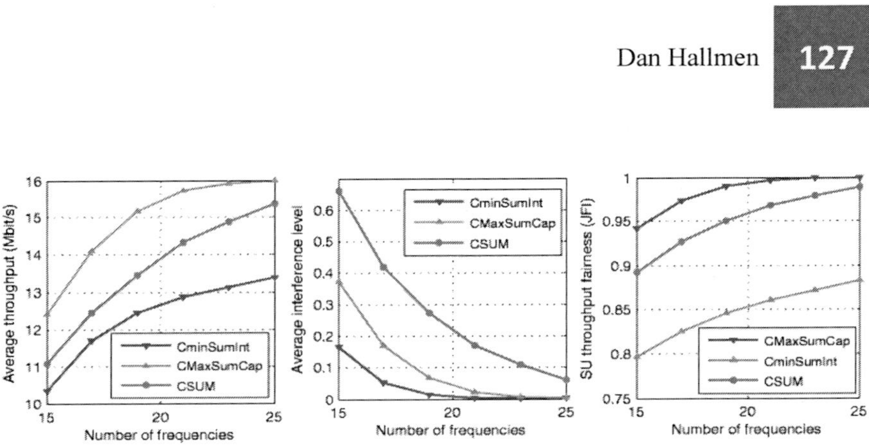

Figure 9 Centralized FA algorithms' performance depending on the number of frequencies (NoPU = 25, NoSU = 40, M = 15–25). The figure shows average throughput per SU, average interference level per SU, and SU throughput fairness depending on number of frequencies

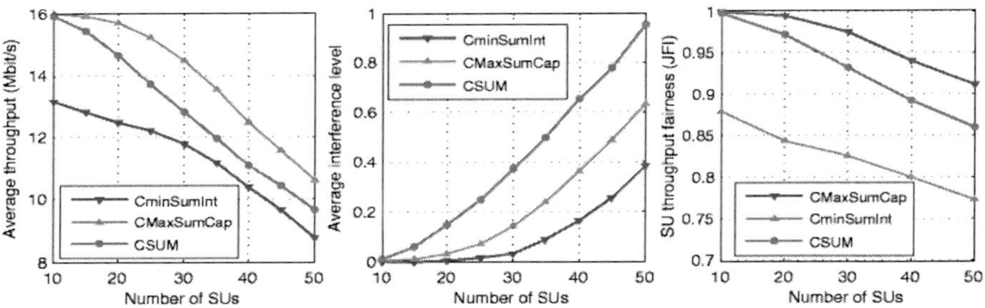

Figure 10 Centralized FA algorithms' performance depending on the number of SUs (NoPU = 25, NoSU = 10–50, M = 15). The figure shows average throughput per SU, average interference level per SU, and SU throughput fairness depending on the number of SUs

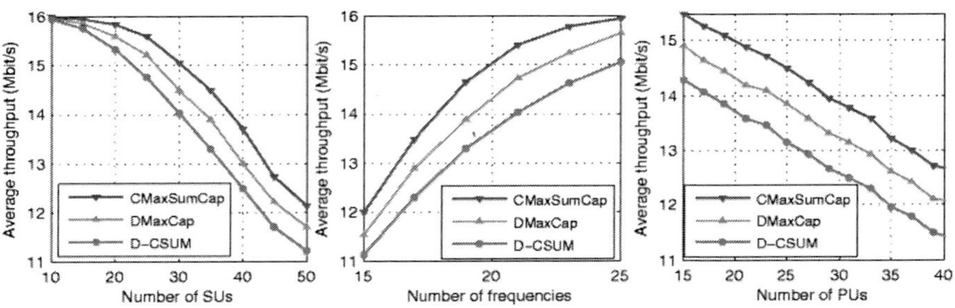

Figure 11 Throughput performance of distributed and centralized FA algorithms (NoPU = 15–40, NoSU = 1–50, M = 15–25). The figure shows average throughput per SU depending on the number of SUs, number of frequencies, and number of PUs

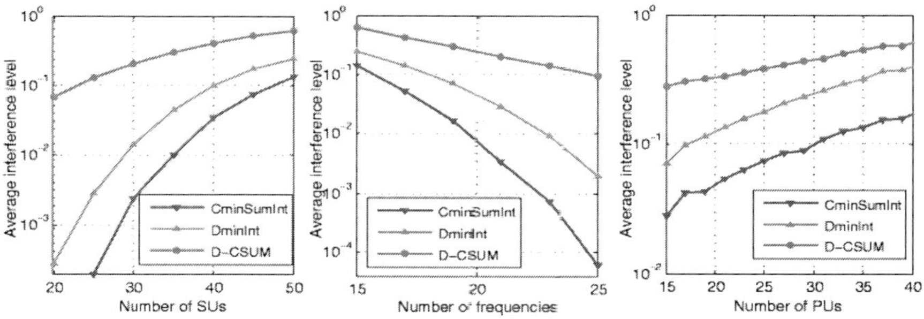

Figure 12 Interference performance of distributed and centralized FA algorithms (NoPU = 15–40, NoSU = 10–50, M = 15–25). The figure shows average interference per SU depending on number of SUs, number of frequencies and number of Pus

Figure 9 shows the average throughput per SU, the average interference per SU, and the throughput fairness depending on the number of the frequencies, for the network with 25 active PUs, 40 SUs, and other simulation parameters according to Table 4. It can be seen that the average throughput per SU steadily increases with the increase of a number of the available frequencies. CMaxSumCap algorithm reaches maximum throughput per SU, which is set at 16 Mbit/s with a frequency pool of 23 channels. In order to achieve an average throughput of 13 Mbit/s per SU, a network with CMaxSumCap algorithm uses 16 frequencies, with CSUM algorithm uses 18 frequencies and with CminSumInt uses 22 frequencies which is 35 % more than the best performing algorithm. The average interference per SU is decreasing with a larger number of the available frequencies, since the same number of SUs is distributed in a larger frequency pool, resulting in a lower number of SUs per frequency and therefore lower mutual interference. If the average mutual interference per SU is limited to 0.1, the CR network with 40 SUs and CminSumInt algorithm can operate with 16 frequencies. The network with CMaxSumCap algorithm needs more than 18 frequencies and network with CSUM algorithm needs at least 23 frequencies. If the number of frequencies is set to 20, it can be observed that the average interference is below 0.05 for CMaxSumCap and CminSumInt algorithms, while it is 0.22 for CSUM algorithm (i.e., 4.5 times larger). This clearly shows the benefits of the interference weighting and categorization on the network performance. A comparison of throughput fairness shows that the algorithms have Jain's fairness index larger than 0.8, meaning that algorithms provide a good fairness of network resources. As the number of available frequencies increases, fairness also increases. From Fig. 9, we conclude that the SU networks using CMaxSumCap and CSUM algorithms have excellent fairness. The network using CminsumInt algorithms has a higher discrepancy in the SU throughput, but nevertheless, it is still satisfactory.

Figure 10 shows the average throughput per SU, the average interference per SU and the throughput fairness depending on the number of SUs in the CR network with 25 active PUs, and the available pool of 15 frequencies. The average throughput decreases, and the interference increases with an increasing number of secondary users in the CR network. Intuitively, when we set the CR network area and available frequencies, while the number of SUs increases, there are more conflicts between SUs, since the distance between the interfering SUs is decreasing. Consequently, there is a smaller choice of the available frequencies, frequencies are more saturated, the SINR per SU is decreasing, and an average throughput per user decreases. If an average interference per SU is set to 0.2, the CR network with 15 frequencies using CSUM algorithm can accommodate 22 SUs, the network using CMaxSumCap algorithm can accommodate 33 SUs, and the network using CminSumInt algorithm can accommodate 42 SUs. Therefore, the results show that network using CMaxSumCap can accommodate almost two times more users than the network using CSUM algorithm. The CR network with 30 SUs using CminSumInt algorithm has a very low average interference of 0.03, comparing to 0.14 for the network using CMaxSumCap algorithm and 0.37 for the network using CSUM algorithm (i.e., 12 times larger). Due to its greedy nature, fairness of all three algorithms in Fig. 10 is decreasing with increasing number of the SUs accessing the network. CMaxSumCap results in Jain's fairness index above 0.9 in whole tuning range, while in networks using CminSumInt algorithm fairness is worse due to larger differences in SU's throughput. Generally, we can conclude that among the studied algorithms, the centralized minimum cumulative network interference algorithm CminSumInt causes the lowest interference levels in all investigated scenarios and most efficiently uses the radio frequency spectrum. The CR networks using CminSumInt algorithm have a low harmful interference footprint, and therefore, they cause low interference pollution of the radio spectrum environment. Having minimized the level of harmful interference to the SUs and limited interference to the primary network, the largest number of the SUs can be accommodated in the available radio spectrum space. Therefore, networks using CminSumInt algorithm provide the most efficient use of the radio spectrum among the studied algorithms. On the other hand, minimizing interference leads to the greater differences of the throughput between the SUs as the JFI score is lesser compared to the other studied algorithms. Centralized interference sensitive maximum throughput frequency assignment algorithm CMaxSumCap has the highest CR network throughput and excellent fairness among the studied algorithms. Considering that, the

CR network using CMaxSumCap algorithm makes the best self-use of the available radio frequency spectrum without causing excessive interference to the PU network. Since the CMaxSumCap algorithm is selecting the frequency with high throughput, causing minimum interference in the selection phase, it is very efficient in frequency usage. CMaxSumCap algorithm causes more interference than CminSumInt algorithm, but it is a well-balanced algorithm, since it distributes throughput more fairly between the users. Both proposed algorithms show significant reduction of the average interference compared to the benchmark binary interference algorithm CSUM. CMaxSumCap algorithm performs better than the benchmark algorithm in throughput value, average interference, and throughput fairness due to the introduction of a co-channel and an adjacent channel interference gradation with edge weights and new saturation metric for dynamic sorting of vertices in the process of frequency assignment.

8.4 Distributed FA algorithms performance

In this section, we evaluate the performance of the distributed FA algorithms in the CR networks, as presented in Figs. 11 and 12. Comparing the distributed algorithms to their centralized counterparts as a benchmark, distributed algorithms are inferior because they perform a local optimization only, with the cooperation among the neighboring SUs, while the centralized algorithms tend to objectivize the whole CR network under their control, resulting in an overall better results. However, distributed algorithms are much faster as shown in Fig. 8, more robust and simpler to implement.

Figure 11 shows an average throughput per SU depending on the number of SUs, the number of frequencies, and the number of PUs for centralized and distributed algorithms. In the distributed algorithms, the average throughput per SU decreases with the increase of the number of PUs or SUs, similarly to centralized algorithms. Also, an average throughput per SU increases with a larger number of available frequencies, due to a lower frequency congestion. An average SU throughput in the network of 30 SUs and with 15 frequencies is 14 Mbit/s for distributed CSUM algorithm (D-CSUM), 14.48 Mbit/s for DMaxCap algorithm, and 15 Mbit/s for CMaxSumCap algorithm. In the environment with 25 active PUs and CR network with 50 SUs, 18 frequencies (17.84) is needed for CMaxSumCap algorithm, 20 frequencies (19.31) for DMaxCap algorithm, and 21 frequencies (20.95) for D-CSUM algorithm, in case of setting 14 Mbit/s for the average throughput. In the CR network with 40 SUs and 19 frequencies assigned to 25 PUs, an average throughput for the CR network using D-

CSUM algorithm is 13.1 Mbit/s, for the CR network using DMaxCap is 13.8 Mbit/s and for the CR network using CMaxSumCap is 14.5 Mbit/s. Comparing an average network throughput per SU, we can conclude that the distributed algorithm achieves 5 % lower average throughput per user compared to a centralized version of the algorithm with the advantage of a faster algorithm convergence and reduced network complexity and communications overhead. The throughput difference between centralized and distributed algorithms is not large, but if we sum over a large number of SUs, the total CR network throughput of the network with distributed frequency assignment is lower than the CR network with the centralized frequency assignment algorithm.

Figure 12 shows an average interference per SU depending on the number of SUs, the number of frequencies, and the number of PUs for centralized and distributed algorithms. Having primary network of 25 PUs using 15 frequencies, CR network utilizing D-CSUM algorithm can operate 23 SUs with an average interference 0.1, CR network utilizing DminInt algorithm can handle 40 SUs, and the CR network utilizing CminSumInt algorithm can operate 47 SUs. The CR network with 50 SUs using a distributed frequency assignment algorithm needs two to three frequencies more than the CR network using a centralized algorithm, which roughly corresponds to 15 % but significantly less than the SU network using D-CSUM algorithm. Increasing the number of PUs in the analyzed area deteriorates the performance of the CR network because the number of available frequencies is reduced due to the fact that SUs cannot operate on the same frequencies as PUs in the interfering area.

From the presented analysis, we can conclude that the distributed algorithms are less efficient in frequency assignment compared to their centralized counterparts. On the other hand, the distributed algorithms are simpler and easier to implement, since they do not require an extensive cooperation and communication between the SUs. Taking into account analyzed and presented distributed frequency assignment algorithms, we can conclude that the distributed minimum interference algorithm DminInt causes the lowest interference levels and most efficiently uses the radio frequency spectrum in all investigated distributed scenarios. Considering a throughput performance, distributed interference sensitive maximum throughput frequency assignment algorithm DMaxCap has the highest CR network throughput and the lowest SUs throughput imbalance among the studied distributed algorithms.

9 CONCLUSIONS

In this paper, we have formulated and addressed the FA problem in the CR networks. The FA ensures that the appropriate frequency is selected to satisfy the requirements of the SUs, while enabling an efficient usage of the radio frequency spectrum. The FA is a crucial function that limits the interference between the CR devices and PUs operating in a heterogeneous and dynamic radio environment. We have treated the FA problem as a graph coloring problem, while introducing the CR specificity. We have introduced the interference susceptibility with a two-layered conflict graph determining the interference potential using co- and adjacent channel edge weights. The protection of the PUs is realized with a local dynamically changing list of blocked channels in the PUs interference range. We have proposed the interference characterization with four categories as an extension to a continuous interference model. As a generalization of our model, we proposed frequency and bandwidth selection instead of channel selection in the FA decision process. We have presented centralized and distributed sequential algorithms that can assign channels to the SU communication links in the CR network with the objective of minimizing network interference (CminSumInt, DminInt) and maximizing network throughput (CMaxSumCap, DMaxCap). The algorithms use dynamic vertex ordering with CR specific novel saturation metrics, which takes into account the frequency limitations due to the PUs activities and interference potential of the already assigned neighboring SUs determining the degree of freedom in selecting the frequency. We showed the effectiveness of our algorithms in the reducing interference and improving the SU network throughput using the network simulation and performance evaluation.

CminSumInt algorithm minimizes the influence of the SU to the other PUs and SUs, resulting in 70 % average reduction of the interference compared to the benchmark CSUM algorithm. CMaxSumCap maximizes the SU network throughput resulting in 15 % increase of the average throughput per SU. CMaxSumCap is a well-balanced algorithm with a good performance in all three performance indicators: throughput, fairness, and interference. The CR networks using the FA algorithms with the interference weighting and categorization can accommodate almost twice as much CR users compared to the networks using algorithms with only binary interference model. Distributed algorithms DminInt and DMaxCap carrying out local optimization provide benefits comparable to a centralized approach. The distributed algorithms reduce the computation complexity and have the faster algorithm convergence and linear increase with the number of SUs. It is shown that the centralized algorithms are more suitable for smaller CR networks, or networks with longer lasting spectrum opportunities, while distributed algorithms are more suitable for the bigger SU networks and more dynamic spectrum environment.

We can conclude that the proposed interference weighting and categorization is beneficial to the CR network performance, since it is possible to quantify the individual interference components and aggregate interference. This approach results in a significantly reduced mutual influence between terminals (2.5–12 times less then with binary interference model) and more efficient spectrum usage. Based on the proposed interference metric, the resource manager can make more knowledgeable and more frequency efficient decisions. The downturn of the interference minimum algorithms is in lower fairness of the throughput. This problem is not present in the maximum capacity algorithms. In future work, the proposed framework can be extended to add a selection of appropriate frequency band and type of the CR spectrum access for multi-band CR operation in heterogeneous environment and to investigate implementation of the fuzzy logic membership function for the interference categorization.

DECLARATIONS

Acknowledgements

The authors would like to thank Dr. Hongjian Sun, the associate editor coordinating the review of this paper and approving it for publication. The authors would also like to express sincere appreciation to the anonymous reviewers and Dr. Ninoslav Majurec from NASA-JPL for their valuable comments which helped in significantly improving this manuscript.

Competing interests

The authors declare that they have no competing interests.

REFERENCES

1. J Mitola, GQ Maguire Jr, Cognitive radio: making software radios more personal. Pers. Commun. IEEE. **6**(4), 13–18 (1999).
2. J Mitola, Cognitive Radio—An Integrated Agent Architecture for Software Defined Radio. PhD dissertation, Royal Institute of Technology (KTH) (2000).
3. S Haykin, Cognitive radio: brain-empowered wireless communications. Sel. Areas Commun. IEEE J. **23**(2), 201–220 (2005).
4. R Murphey, in Encyclopedia of Optimization. vol. 1, 2nd ed., ed. by CA Floudas, PM Pardalos. Frequency assignment problem (Springer Science & Business MediaNew York, 2008), pp. 1097–1101. ISBN: 978-0-387-74760-6.
5. KI Aardal, SP Van Hoesel, AM Koster, C Mannino, A Sassano, Models and solution techniques for frequency assignment problems. Ann. Oper. Res. **153**(1), 79–129 (2007).View ArticleMathSciNetMATH
6. R Beutler, Frequency assignment and network planning for digital terrestrial broadcasting systems (Springer Science & Business Media, New York, 2004). ISBN: 978-1-4020-7872-9.
7. IF Akyildiz, WY Lee, MC Vuran, S Mohanty, A survey on spectrum management in cognitive radio networks. Commun. Magazine, IEEE. **46**(4), 40–48 (2008).
8. G Salami, O Durowoju, A Attar, O Holland, R Tafazolli, H Aghvami, A comparison between the centralized and distributed approaches for spectrum management. Commun. Surv. Tutorials, IEEE. **13**(2), 274–290 (2011).
9. R Zhang, YC Liang, S Cui, Dynamic resource allocation in cognitive radio networks. Signal Proc. Mag. IEEE. **27**(3), 102–114 (2010).View ArticleMathSciNet
10. EZ Tragos, S Zeadally, AG Fragkiadakis, VA Siris, Spectrum assignment in cognitive radio networks: a comprehensive survey. Commun. Surv. Tutorials, IEEE. **15**(3), 1108–1135 (2013).
11. WY Lee, IF Akyldiz, A spectrum decision framework for cognitive radio networks. Mobile Comput. IEEE Trans. **10**(2), 161–174 (2011).

12. AT Hoang, YC Liang, MH Islam, Power control and channel allocation in cognitive radio networks with primary users' cooperation. Mobile Comput. IEEE Trans. **9**(3), 348–360 (2010).

13. C Peng, H Zheng, BY Zhao, Utilization and fairness in spectrum assignment for opportunistic spectrum access. Mobile Netw. Appl. **11**(4), 555–576 (2006).

14. J Wu, Y Dai, Y Zhao, Effective channel assignments in cognitive radio networks. Comput. Commun. **36**(4), 411–420 (2013).

15. H Xu, B Li, Resource allocation with flexible channel cooperation in cognitive radio networks. Mobile Comput. IEEE Trans. **12**(5), 957–970 (2013).

16. M Vishram, LC Tong, C Syin, in Region 10 Symposium, 2014 IEEE. List multi-coloring based fair channel allocation policy for self coexistence in cognitive radio networks with QoS provisioning (IEEEKuala Lumpur, 2014), pp. 99–104. ISBN: 978-1-4799-2028-0.

17. S Anand, S Sengupta, R Chandramouli, MAximum SPECTrum packing: a distributed opportunistic channel acquisition mechanism in dynamic spectrum access networks. IET commun. **6**(8), 872–882 (2012).

18. P Hu, J Ye, F Zhang, S Deng, C Wang, W Wang, in Vehicular Technology Conference (VTC Fall), 2012 IEEE. Downlink resource management based on cross-cognition and graph coloring in cognitive radio femtocell networks (IEEEQuebec City, 2012), pp. 1–5. ISBN: 978-1-4673-1880-8.

19. AP Subramanian, H Gupta, SR Das, J Cao, Minimum interference channel assignment in multiradio wireless mesh networks. Mobile Comput. IEEE Trans. **7**(12), 1459–1473 (2008).

20. N Nie, C Comaniciu, P Agrawal, in Wireless Communications. A game theoretic approach to interference management in cognitive networks (Springer Science & Business MediaNew York, 2007), pp. 199–219. ISBN: 978-0-387-37269-3.

21. RM Karp, in Complexity of computer computations. vol. 1, 1st ed, ed. by RE Miller, JW Thatcher. Reducibility among combinatorial problems (Plenum PressNew York, 1972), pp. 85–103.

22. D Brélaz, New methods to color the vertices of a graph. Commun. ACM. **22**(4), 251–256 (1979).

23. Z Tabakovic, S Grgic, M Grgic, in Systems, Signals and Image Processing, 2009. IWSSIP 2009. 16th international conference on. Fuzzy logic power control in cognitive radio (IEEEChalkida, 2009), pp. 1–5. ISBN: 978-1-4244-4530-1.

24. F Huang, W Wang, H Luo, G Yu, Z Zhang, in Vehicular technology conference (VTC 2010-Spring), 2010 IEEE 71st. Prediction-based spectrum aggregation with hardware limitation in cognitive radio networks (IEEETaipei, 2010), pp. 1–5. ISBN:978-1-4244-2518-1.

25. ST Chung, AJ Goldsmith, Degrees of freedom in adaptive modulation: a unified view. Commun. IEEE Trans. **49**(9), 1561–1571 (2001)

26. R Jain, DM Chiu, WR Hawe, A quantitative measure of fairness and discrimination for resource allocation in shared computer system (Digital Equipment Corporation, Hudson, 1984).

6

CHAPTER

COGNITIVE RADIO FOR VEHICULAR *AD HOC* NETWORKS (CR-VANETS): APPROACHES AND CHALLENGES

Kamal Deep Singh1* , Priyanka Rawat2 and Jean-Marie Bonnin1

1Department RSM, Telecom Bretagne, 35510 Cesson Sevigne, France.

2INRIA Lille - Nord Europe, Univ. Lille 1, CNRS UMR 8022, 59650 Lille, France.

ABSTRACT

With growing interest in using cognitive radio (CR) technology in wireless communication systems for vehicles, it is envisioned that future vehicles will be CR-enabled. This paper discusses CR technologies for vehicular networks aimed at improving vehicular communication efficiency. CR for vehicular networks has the potential of becoming a killer CR application in the future due to a huge consumer market for vehicular communications. This paper surveys novel approaches and discusses research challenges related to the use of cognitive radio technology in vehicular *ad hoc* networks. We review how CR technologies such as dynamic spectrum access, adaptive software-defined radios, and cooperative communications will enhance vehicular communications and, hence, present the potential of transforming vehicle communication in terms of efficiency and safety. Our work is different from existing works in that we provide recent advances and open research directions on applying cognitive radio in vehicular *ad hoc* networks (CR-VANETs) focusing on architecture, machine learning, cooperation, reprogrammability, and spectrum management as well as QoE optimization for infotainment applications. A taxonomy of recent advances in cognitive radio for vehicular networks is also provided. In addition, several challenges and requirements have been identified. The research on applying CR in vehicular networks is still in its early stage, and there are not many experimental platforms due to their complex setup and requirements. Some related testbeds and research projects are provided at the end.

Keywords: CR-VANETs CRVs Dynamic spectrum allocation IEEE 802.11p Intelligent transportation systems Survey Vehicular communication V2I V2R V2V

1 INTRODUCTION

Increasing number of vehicles on the road has brought focus on improving road safety as well as in-vehicle entertainment. In tune with this demand, we are witnessing a rise in development of new applications and services for vehicular environments. Some common examples include applications for collision avoidance, safety and traffic monitoring, multimedia streaming, data collection for smart cities in synergy with wireless sensor networks [1], vehicle-to-vehicle communication (V2V), etc. Consequently, vehicular *ad hoc* network (VANET) has emerged as a new technology that can support such emerging vehicular applications.

A VANET is defined as a spontaneous *ad hoc* network formed over vehicles moving on the road. Such a network can be formed between vehicles with V2V communication or between vehicles and infrastructure with vehicle-to-infrastructure (V2I) communication. Such VANETs in which vehicles can communicate with each other and also with roadside infrastructure provide a means to improve road safety by enabling a number of potential applications for driver assistance, collision warning, traffic information, and monitoring [2–6]. The availability of various applications will improve road safety and vehicular environment.

Dedicated short-range communication (DSRC) [7–10] is a generic name for short-range, point-to-point communication. It is also the name of the older technology mainly used for vehicle to road communication (e.g., toll gate). The channels reserved worldwide in the 5.9 GHz band for such communications are known as DSRC channels. The IEEE 1609 - Dedicated Short-Range Communication Working Group is developing standards for wireless access in vehicular environments (WAVE) and the communication

is based on IEEE 802.11p [11, 12], which is an amendment to IEEE 802.11 standard in order to support communication in dynamic vehicular environments. IEEE 802.11p standardizes the communication aspects related to physical (PHY) and media access control (MAC).

Currently, the US Federal Communications Commission (FCC) has allocated 75 MHz and the European Telecommunications Standards Institute (ETSI) allocated 30 MHz of spectrum in 5.9 GHz band for the deployment of Intelligent Transportation Systems (ITS) services [13, 14]. However, a significant rise in vehicular applications, especially in urban environments, with several vehicles, may lead to overcrowding of the band and thereby resulting in degraded vehicular communication efficiency for safety applications, as pointed out in [15]. Moreover, not only safety applications, but also growing demand and usage of in-car entertainment and information systems comprising bandwidth demanding multimedia applications (e.g., video streaming) will lead to congested vehicular networks and spectrum scarcity for IEEE 802.11p-based vehicular applications [15].

In view of this, using cognitive radio (CR) technology in VANETs will enable more efficient radio spectrum usage and, in turn, improve vehicular communication efficiency. Figure 1 illustrates a V2V scenario with cognitive radio which allows for opportunistic spectrum usage, along with the utilization of DSRC channels. Spectrum utilization measurements over the years have indicated an inefficient spectrum allocation and usage with many unused or underused licensed bands over different space and time, for example, spectrum bands for TV broadcasting, resulting in considerable spectrum wastage [16–20]. Consequently, regulatory agencies, such as FCC [21] in the USA that regulates spectrum allocations, have now opened the licensed bands to unlicensed/secondary user (SU) through the use of CR [22–27] in order to enable a more efficient use of the spectrum bands. CR is an emerging technology to improve spectrum usage and alleviate spectrum scarcity by exploiting underused spectral resources through opportunistic spectrum access.

Radio cognition (RC) together with dynamic spectrum access (DSA) strategies present promising approaches that allow unlicensed/secondary users to opportunistically capture and use the spatio-temporally available licensed spectrum holes as long as licensed/primary user (PU) are not perturbed [25]. Though many applications of CR have been proposed for wireless mesh and in *ad hoc* and cellular networks, employing CR in vehicular (CRV) environments is a relatively novel research subject and some works in that direction have started to appear recently [28–35].

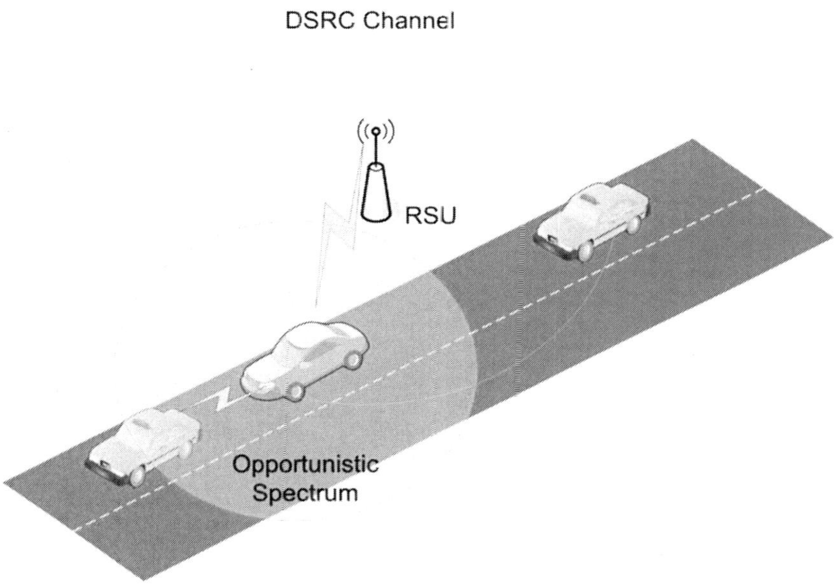

Figure 1 OPPORTUNISTIC SPECTRUM USE FOR V2V. RSU, roadside unit; DSRC, dedicated short-range communication.

Vehicular networks and vehicular communication can benefit from opportunistic spectrum usage with CR technology [36]. Cognitive radio for vehicular *Ad hoc* Networks (CR-VANETs) is a fast-emerging application area of CR technology [28]. CR-enabled vehicles can use additional spectrum opportunities in TV bands according to the quality-of-service (QoS) requirements of the applications. However, the research solutions proposed for general-purpose CR networks cannot be directly applied to CR-VANETs. This is because the unique features of vehicular environment such as high mobility and cooperation opportunities need to be taken into account while designing the spectrum management functions for CR-VANETs. Unlike static CR scenarios, in CR-VANETs, multiple cooperating vehicles (during busy hours) can exchange spectrum information to get information on the spectrum availability. This enables other vehicles to know the spectrum characteristics of the road to be traversed in advance and they can take proactive adaptive measures. In general, CR-VANETs will improve the performance of existing and emerging vehicular applications such as V2V communication, entertainment and Information Systems, and public safety communication.

In this paper, we focus on CR-VANETs. Section 2 provides some background and challenges on vehicular networks followed by an overview of cognitive radio technology. Section 3 discusses the motivations of using cognitive radio technology in vehicular networks. A survey of recent advances in CR-VANETs is provided in Section 4, and Section 5 describes some related projects and testbeds. Finally, some open issues which need further attention from the research community are discussed in Section 6 followed by conclusions in Section 7.

2 BACKGROUND

This section provides some background and challenges on vehicular networks followed by an overview of cognitive radio technology.

2.1 Vehicular communications

Modern vehicles are making inroads in the market. These vehicles are not only equipped with global positioning system (GPS) and navigation systems, but also more advanced features such as environmental awareness to prevent vehicle collisions, multimedia systems, and integrated wireless access systems to improve vehicle performance and user experience. In addition, there is much interest in improving the efficiency of vehicular communications. For this purpose, ITS aim at improving safety, reliability, efficiency, and quality of transport infrastructure and vehicles through the use of information and communication technologies (ICT). Additionally, ITS focus on providing sustainable and affordable transportation by designing advanced applications and services to optimize transportation times and energy consumption.

ITS support different communication scenarios including all types of communications in vehicles, between vehicles, as well as between vehicles and roadside infrastructure. Figure 2 shows the taxonomy of vehicular communications. Vehicular communications can be classified into V2V and V2I. V2I further includes vehicle-to-roadside (V2R) communication and communication using cellular networks. In V2V communications, a vehicular *ad hoc* network (VANET) is formed among vehicles for exchanging information e.g., safety information. In V2R, information is exchanged between the roadside unit (RSU) and the onboard unit (OBU) of a vehicle. In V2I, information is exchanged between the RSU, or possibly a cellular network, and OBU.

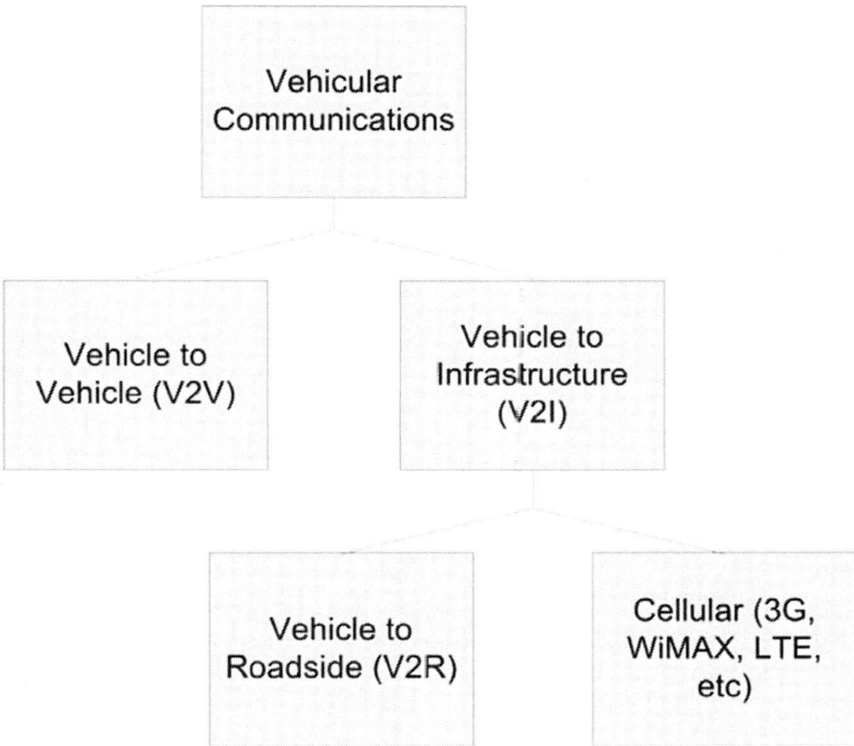

Figure 2 TAXONOMY OF VEHICULAR COMMUNICATIONS.

V2V communications are relatively more complex and challenging as compared to V2R and V2I communications. The V2I communications are more to exchange information with the centralized servers somewhere in the Internet through various access networks and technologies. Currently, there is continuing effort by cellular network operators to enhance the network capacity that will improve the cellular network support for V2I communications. The V2R communications that are presently deployed are mainly short-distance one-to-one communications, for example, toll collection, whereas the V2V communications are expected to be mostly *ad hoc* communications with transmission range up to several 100 m. However, currently, the V2V communications are not yet *ad hoc*-based. There are mainly one-hop communications for unicast and multicast. Standardization bodies and researchers are working on multihop for geo-casting and V2V safety applications, but there are still strong scalability concerns. The transmission range commonly ensured is 300 m in urban area and about 1 km in free space.

2.1.1 Standardization

The process of standardization for wireless access technology to provide connectivity in VANETs is a work in progress. Older DSRC standards were developed for V2V or V2R for safety and other services, such as fee collection in toll plazas. As mentioned earlier, new DSRC is mostly used as a generic name for short-range, point-to-point communication. It is also used to name the worldwide channels in the 5.9 GHz band, which are reserved for vehicular communications.

Currently, several standards and technologies are available including cellular (2G/3G/4G) technology and IEEE 802.11p standard that can be used for high-speed vehicular communication. The challenge is to make different technologies and standards interoperable. Cellular (2G/3G/4G) technology provides good coverage and sufficient security, but it is relatively costly.

The IEEE 802.11p standard will support high-speed vehicular communication at speeds up to 200 km/h in 5.9 GHz band, within the USA. It is a link layer technology based on the former IEEE 802.11a amendment. Today, the IEEE 802.11p amendment has been integrated in the new version of IEEE 802.11 standards. It is mainly an adaptation to support efficient infrastructure-less communications (without any prior attachment). In Europe, this technology is promoted by the vehicular industry as a standard for V2V communications and thus, deployment cost is expected to be comparatively less because of large-scale production. IEEE 802.11p defines MAC and PHY operations. IEEE also standardized a family of IEEE 1609 standards for WAVE that build on top of IEEE 802.11p. The architecture is shown in Figure 3 and it defines various functionalities related to applications, networking, management, communication, as well as security. A *multichannel operation* for IEEE 802.11p is also defined which includes a control channel and six service channels. The management plane defines MAC layer management entity (MLME) and physical layer management entity (PLME) for MAC and PHY layers, respectively. For networking, some applications such as infotainment can use TCP/IPv6 or UDP/IPv6, whereas WAVE short message protocol (WSMP) is additionally defined for applications which have strict latency requirements, i.e., safety applications. Nevertheless, IEEE 1609 is not generic, as it is based only on IEEE 802.11p.

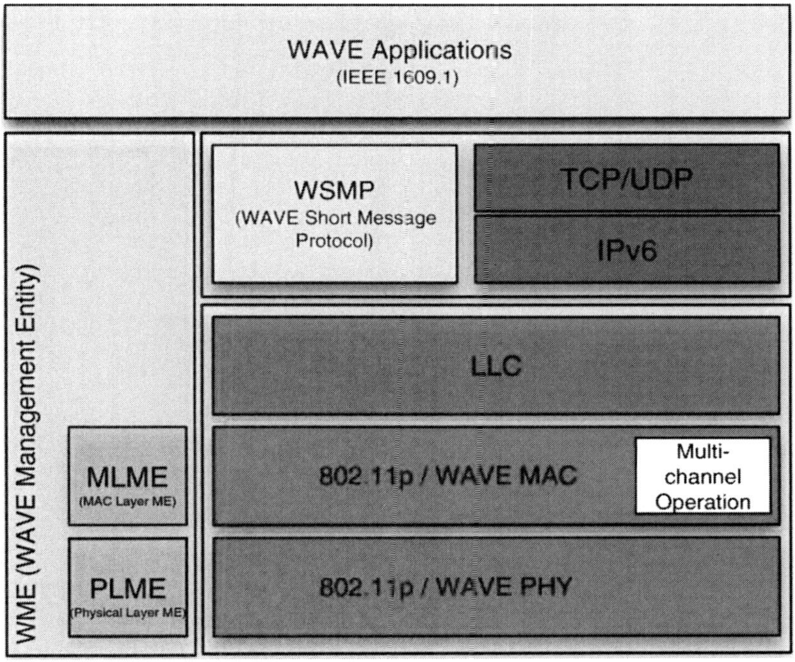

Figure 3 IEEE 1609 WAVE ARCHITECTURE.

The International Standards Organization (ISO) is developing a generic architecture for ITS [14]. Communication access for land mobiles (CALM) refers to a family of standards developed at ISO TC204 Working Group 16 for cooperative ITS. A simplified version of the ITS reference architecture [37] for CALM is shown in Figure 4. Today, different heterogeneous standards exist as a result of *silo-application*-driven approach. The ITS reference architecture defines a generic protocol stack which aims at enabling convergence. The communication media layer, shown in the figure, considers various media technologies and standards, including cellular (GSM/GPRS, UMTS), older DSRC standards, IEEE 802.11 standards, M5 [38] alias IEEE 802.11p, bluetooth, and infrared. The networking and transport layer considers IPv6, network mobility (NEMO), geo-routing, etc. The facilities layer provides support for application, information, and session. A management plane and a security plane are connected to all the layers for cross-layer functionalities.

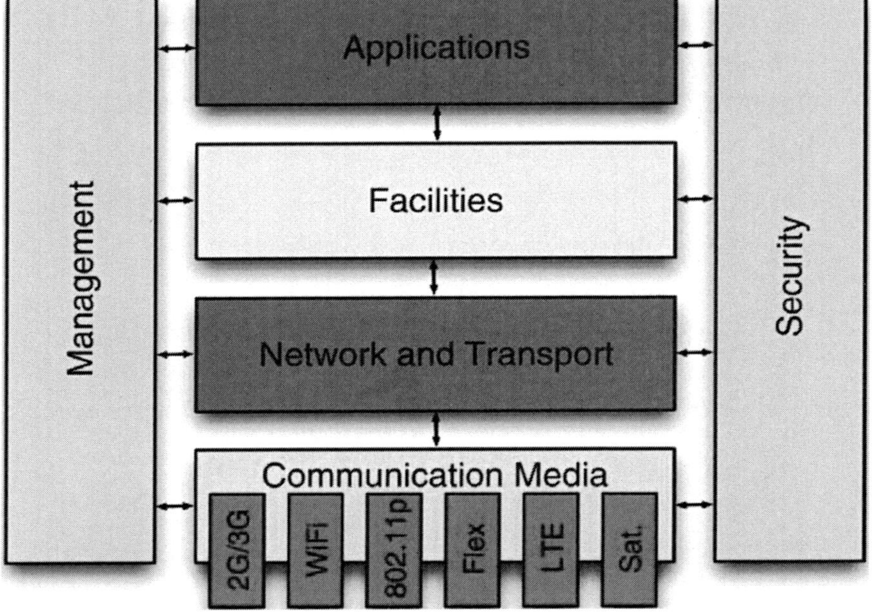

Figure 4 ITS REFERENCE ARCHITECTURE.

Additionally, some harmonization efforts at ISO include the incorporation of IEEE 802.11p and harmonization of CALM-FAST [39, 40] (used for short messages having real-time constraints) with IEEE 1609. M5 [38] refers to the adaptation of IEEE 802.11p to the ISO CALM architecture, and G5 [41] is the adaptation of IEEE 802.11p to European constraints by ETSI. Thus, IEEE 802.11p is sometimes referred to as IEEE 802.11p alias M5 alias G5.

2.1.2 Spectrum policy and regulations

Several portions of the radio spectrum are regulated by the governments or regulatory bodies for an efficient use of the limited radio spectrum. The increasing use of wireless communications systems dedicated to vehicles will require spectrum availability for V2V communication system. Consequently, FCC has allocated 75 MHz of radio spectrum at 5.9 GHz for V2V and V2I in the USA. However, due to the unavailability of a continuous spectrum of 75 MHz in DSRC band in Europe, Car2Car Communication Consortium (C2C-CC), a nonprofit, industry-driven European organization, has proposed to allocate 2 × 10 MHz for primary use of safety critical applications at 5.9 GHz range (5.875 to 5.925 GHz). Since this band is used as control channel in the USA, its allocation in Europe will allow for worldwide harmonization.

Recently (since 2012), Japan has allocated 10 MHz in 700 MHz band for ITS because these bands are appropriate for vehicular communications. Radio wave transmission through obstacles is much easier in the 700 MHz bands [42], whereas radio waves in 5.8 to 5.9 GHz are good for short-range communications, e.g., highway toll collection.

Presently, the number of wireless-enabled vehicles is very low and their spectral bandwidth requirements are low as well. However, the increasing number of wireless-enabled vehicles, vehicular communication applications, and high data rate traffic flows will lead to more and more V2V and V2I information exchanges facilitated by wireless communications. ITS will more and more use different wireless access technologies to improve the efficiency and safety of vehicular communication and transportation. In general, different types of ITS depend on radio services for communication and use specially designed technologies. However, if we consider IEEE 802.11p, then it is limited in terms of capacity because it is based on IEEE 802.11a, a quite old version of Wi-Fi (1999).

V2V and V2R are mainly assured by dedicated communication technology in the ITS frequency bands (5.8 to 5.9 GHz). There is recently a strong pressure on these bands since some companies (Qualcomm, Cisco, etc.) have asked the federal agency to release a part of these bands for more general usages, for example Wi-Fi. Consequently, in the future, these bands could be shared between ITS and other applications, which may lead to crowding of the bands. In addition, the growing demand for vehicular applications (V2V applications) will result in scarcity of spectrum dedicated to vehicle communications. Moreover, vehicle communications-related specific aspects such as high mobility of the vehicles and fast changes in vehicular network topologies add to the challenges in having efficient vehicular communications (Table 1). Therefore, novel techniques and approaches are required to address the problem of spectrum scarcity in VANETs allowing an efficient use of radio spectrum.

Table 1 CHALLENGES AND REQUIREMENTS FOR CR-VANETS

Challenges	Required mechanisms
Reliable delivery of safety messages	QoS support
Highly dynamic topologies	Algorithms with fast convergence time
Scarce spectrum	Efficient use of spectrum
Newer and several technologies	Reprogrammability
Varied environment and spectrum availability	Machine learning algorithms

With this view, our paper discusses cognitive radio technology for vehicular networks in order to improve vehicular communication efficiency. Cognitive radio techniques can be used in the roadside units (RSUs) as well as the onboard units (OBUs) in a vehicle. CR will enable them to adapt their transmissions to deal with the fast changes in the radio environment. CR-enabled vehicles will also be able to adaptively select a radio access technology, among many, for communicating with different types of RSUs.

2.2 VANET challenges

Here we describe the unique characteristics of VANETs and identify some major issues. Deploying a vehicular networking system requires addressing several challenges posed by the unique characteristics and requirements of vehicular communications [43].

2.2.1 Mobility and dynamic network topology

High mobility (for instance, 100 to 200 km/h) of vehicles makes the topology of VANETs very dynamic resulting in very short lived vehicular communication links. Additionally, vehicular density keeps varying from sparse to dense, and high mobility in sparse areas may cause fragmentation problem for VANET, which, in turn, will result in network unreachability for some nodes. Further, high speeds can deteriorate signal due to Doppler and fast fading. These factors can degrade the performance of applications that have QoS requirements in terms of high reliability, low latency, etc.

2.2.2 Distributed *ad hoc* coordination and one-channel vs. multiple-channel paradigm

In V2I communications, the fixed roadside units can serve as coordinators. However, V2V communications are expected to be self-organizing and to function with or without roadside assisting units. Consequently, authors in [3] argue that one-channel paradigm, with a single shared control channel, is a good solution for V2V communications in the absence of central coordination, considering that various applications will be broadcasting messages to many neighboring vehicles. However, one-channel paradigm comes with the problem of hidden terminal and poses difficult requirements on the design of MAC protocol for V2V communications. Though IEEE 802.11 carrier sense multiple access (CSMA)-based MAC is good for V2V communications, its performance degrades in the presence of high number ofusers.

Moreover, if we reach a larger number of vehicles, the dissemination protocols could lead to a larger overhead. Additionally, a high data traffic density may lead to channel congestion, e.g., in case of an accident and consequent outburst of messages. Multiple-channel paradigm can be a potential solution for such scenarios where instant sharing of message is required between vehicles and thereby reducing congestion on common control channels (CCC). Currently, the approach that is in use is to let all vehicles synchronize to a global time reference and alternate between a common control channel and separate service channels every 100 ms. However, this approach is not efficient [44].

2.2.3 Routing issues

Conventional routing protocols are not suitable for VANETs due to their specific network characteristics, e.g., varying network topology and frequent disconnections. Some of the VANETs' routing algorithms can be categorized as opportunistic forwarding [45], trajectory-based forwarding [46], and geographic forwarding [47]. Opportunistic forwarding algorithms (carrying and forwarding a message whenever given the opportunity) are useful in scenarios with frequent disconnections and can be combined with other approaches that use trajectory-based or geographic forwarding. Geographic forwarding algorithms forward packets towards the destination as a function of its geographical location. This routing approach is scalable but not efficient for handling dead ends and voids. Trajectory routing can be the most suitable message forwarding algorithm for VANETs as it considers the road infrastructure as an overlay directed graph, with intersections as graph nodes and roads as graph edges allowing messages to move in predefined trajectories. Moreover, some recent opportunistic approaches

for delay-tolerant applications exploit social networking analyses to forward packets. The idea with social-based forwarding is to forward a packet to a node which has a high chance of meeting the destination node in near future.

2.2.4 Privacy, security, and safety

Privacy and security issues are highly important in VANETs due to potential threats to traffic flow and human lives by any malicious attempt, for example, fake messages leading to traffic disruption and fatal accidents. Some of the security and privacy issues related to ITS have been discussed in [48–50].

Security protocols for vehicular networks should take into account their specific characteristics such as high mobility and requirements such as trust (vehicles should be able to trust the received messages), resiliency (for interference), and efficiency (e.g., ability to authenticate message in real time). In addition, privacy concerns include preserving anonymity so as to prevent tracking or identification of vehicle for non-trusted parties based on vehicular communication [43]. Nevertheless, such security mechanisms generally come at the cost of degraded communication performance [51].

2.3 Cognitive radio and software-defined radio

With the increasing demand for spectral efficiency, cognitive radio has emerged as a very active research area for wireless communications and networks' research community in recent years. CR [26] is an emerging technology that enhances the performance of existing radio by integrating artificial intelligence (AI) with software-defined radios (SDRs). Unlike conventional radios in which most of the components are implemented in the hardware, SDRs [24] are radios that use software implementations for some functionalities enabling flexible radio operation. These radios are reconfigurable, and hence the need to modify existing hardware isreduced.

Though several definitions have been provided to describe CR, the commonly used definitions in literature are given by Joseph Mitola and Simon Haykin. Joseph Mitola described CRs [22–24] as intelligent radios that can autonomously make decisions using gathered information about the radio frequency (RF) environment through model-based reasoning and can also learn and plan according to their past experience. Afterwards, Simon Haykin defined CR as an intelligent wireless communication system capable of being aware of its environment, learning, and adaptively changing its operating parameters in real time for providing reliable communication and efficient utilization of the radio spectrum [25].

As mentioned previously, a cognitive radio can be built on software-defined radio (SDR) that makes it reconfigurable by converging the two key technologies: software-defined radio and artificial intelligence [52–57]. While the field of CR is independent from SDR, CR has its origin in SDR as an enabling technology to support its implementation.

In the context of VANETs, with increasing number of vehicles equipped with wireless communication systems, vehicular communication will require faster connectivity and thus, wireless spectrum will have to be adapted to the new requirements (of bandwidth). Using CR and SDR will prevent the need to implement hardware upgrades with emergence of new protocols in future. SDR enables 'reconfigurable systems' for wireless networks. CR allows SDR to determine which mode of operation and parameters to use. In general, CR and SDR will allow CR-enabled vehicles to search for the best frequency-based predetermined parameters.

2.3.1 Cognition cycle

Here we first describe the two main features of CR: cognitive capability and reconfigurability. Then, we briefly discuss the concept of cognition cycle of CR as well as some specifics related to CR-VANETs.

A CR-enabled device adapts its operational parameters as a function of its environment [18]. CR components are mainly radio, sensor, knowledge database, learning engine, optimization tools, and a reasoning engine. CR has cognitive as well as reconfigurability capabilities [25, 26, 58]. Cognitive capability allows CR to sense and gather information (e.g., different signals and their modulation types, noise, transmission power, etc.) from its environment and, for example, secondary users can identify the best available spectrum. The reconfigurability features of CR allow it to optimally adapt the operational parameters as a function of the sensed information. CR systems involve PU and SU of the spectrum; primary users are license holders, while secondary users seek to opportunistically use the spectrum through CR when the primary users are idle.

The *cognition cycle* of CR consists of multiple phases: Observe, Analyze, Reason, and Act. The goal is to detect available spectrum, select the best spectrum, select the best operational parameters, coordinate the spectrum access with other users, reconfigure the operational parameters, and vacate the frequency when a primary user appears

A spectrum hole refers to a portion of spectrum not being used by the primary/licensed user at a particular place and time. It is detected through spectrum sensing and signal detection techniques. The SUs opportunistically access the spectrum if the sensed portion of spectrum is found empty. The secondary user can opportunistically use different spectrum bands by adaptive spectrum switching through spectrum hand-offs. However, the secondary user is responsible to detect arrival of any primary user. If a primary user is detected, then it should vacate the licensed portion of the spectrum in order to prevent interference.

A possible cognition cycle for CR-VANETs in shown in Figure 5 having the four stages: Observe, Analysis, Reasoning, and Act with some customized functionalities for vehicular networks. The Observe stage consists of sensing as well as the consideration of location, policies, and application QoS needs. In the Analysis and Reasoning stages, the system performance and the radio environment are analyzed, similarities from the past are considered, and optimal parameters as well as optimal spectrum strategies are determined. The reconfiguration and optimal adaptation is finally done in the Act stage.

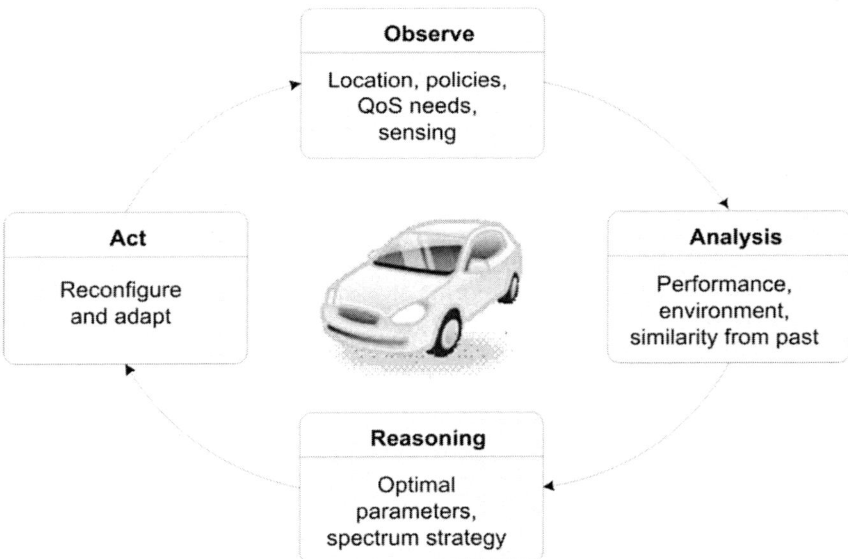

Figure 5 CR-VANETS COGNITION CYCLE.

The CR-VANETs cognition cycle has a lot of similarities with that of CR, but some differences are due to the nature of high-mobility vehicular environment. Most of the vehicles will be equipped with navigational systems, and hence location as well as mobility prediction based on the current direction of movement can be used for cognition. The use of knowledge database becomes more relevant as a vehicle may pass through the same location at approximately the same time of the day. Thus, past experiences can be combined with the current location for optimizing the operational parameters. Finally, it should be noted that the requirements for CR-VANETs are more strict in terms of faster adaptability and faster running time of the cognition cycle due to the dynamic nature of the vehicular environment.

3 MOTIVATIONS FOR CR-VANETS

Currently, high-speed modern vehicles and a large number of cars move on the road every day. ITS [13] aim to develop advanced applications for improving safety and efficiency of road infrastructure and transportation. This will enable users to utilize transport networks in a smarter and safer way.

Presently, the primary objective of vehicle industry is to improve the traveling experience of users by enhancing vehicular communication capabilities with better safety and efficiency, as well as Internet access and infotainment applications.

The number of vehicles on road is always increasing as well as the demand for intervehicle communications. However, ITS and vehicle communications need to overcome several technical challenges. Protocols and applications designed for vehicular communications should consider various important factors such as communication infrastructure, road infrastructure, vehicle density, user demands and types of vehicular networks, as well as available wireless spectrum. Many envisioned applications will need uninterrupted and reliable connectivity and this can be challenging in high-speed vehicular scenarios. New protocols are needed that can deal with rapidly changing environment and are fault tolerant especially for applications related to safety. Overall requirements for such applications are QoS support, adaptability to fast changing environment, robustness, and additional bandwidth to deal with congestion and high bandwidth requirements of some applications such as video streaming. We argue that the cognitive radio technology can help in answering some requirements, and the motivations leading to CR-VANETs are provided in the remaining part of this section.

3.1 QoS requirements

QoS support is important for applications such as related to safety. From the point of view of radio technology, satisfying QoS guarantees is easier when there is enough bandwidth, which can be traded with QoS guarantees such as low delay or high reliability. Additionally, there should be mechanisms to protect important flows from lower priority flows. We discuss the issues related to bandwidth and flow priority in the following text.

3.1.1 Bandwidth scarcity and congestion

Wireless communication is popular as ever and the demand for more bandwidth and spectrum is ever increasing. In addition, applications like video streaming are becoming popular, which, in turn, consume high bandwidth and can cause congestion. Similar problems of bandwidth scarcity and congestion will impact vehicular communications with growing demands. Recently, many car manufactures have started to provide connectivity features in their cars enabling Internet connectivity through cellular networks (that provide mobile voice and data connections) and other wireless access technologies for accessing various applications. However, the performance of cellular networks suffers in urban areas (large cities) due to congestion of cellular spectrum while in rural areas, cellular coverage is either absent or low. Moreover, as mentioned previously, for dedicated communication, ITS are based on IEEE 802.11p, which is based on an old version of Wi-Fi, namely, IEEE 802.11a, and thus, it has limited capacity. Presently, there is a lack of applications for intervehicle communications, which will change in the future. Car manufacturers look unlikely to buy spectrum. Additionally, note that spectrum is a scarce natural resource and should be used efficiently. Thus, opportunistic spectrum access is an attractive solution.

3.1.2 Offloading lower priority flows

Many applications are envisioned for VANETs such as infotainment. Some of these video applications, for example, peer-to-peer (P2P) video sharing and multimedia advertisements can consume a lot of bandwidth and can cause collision with other flows. One idea is to use the main 802.11p spectrum for important flows and use additional bandwidth available with cognitive radio, opportunistic spectrum (see Figure 1) for lower priority video flows, and P2P traffic.

3.2 Resiliency

The flexibility and agility offered by cognitive radio is very useful for resilient communications in vehicular scenarios. In case of emergency situations, CR can be reconfigured in real time to operate in emergency mode by focusing on minimizing bit error rate (BER) and avoiding interference [59]. This property is very useful for ITS safety applications, and with CR functionalities, some objectives like bandwidth maximization or power minimization can be flexibly traded for more resiliency.

3.3 More spectrum holes on highway

In many cases, highways are open spaces and there is a high chance of finding a spectrum hole that can be accessed opportunistically. This is unlike downtown and urban areas where chances of finding spectrum holes can be low due to high population. In fact some experiments have been done [29] which show free abundant spectrum available for opportunistic use on highways. Thus, cognitive radio technology is very attractive as it can exploit such spectrum availability using opportunistic spectrum access. This, in turn, can answer some of the bandwidth and congestion problems discussed above.

3.4 Sufficient space and power supply in vehicles

Some of the advanced cognitive radio capabilities come at a price in terms of bigger size of onboard units. Some functionalities can also consume energy. However, vehicles have sufficient space and power supply and are not limited by them unlike the case with smart phones and other highly portable devices. Cost can still be a factor, but performance vs. cost trade-off can be exploited. Cost can also be reduced with the help of mass production, by optimally designing an OBU with cognitive radio capabilities.

3.5 Reprogrammable vehicular telematics

Every 2 to 3 years, a new communication standard is being proposed, such as DVB-H, DVB-T2, WiMaX, 802.11p, LTE, and HSDPA. Some of these standards face the risk of becoming a commercial failure such as was the case with DVB-H. Moreover, different countries have different regulations related to spectrum usages, transmit power limits, etc. Thus, global manufacturers of vehicles face a dilemma on which technology to deploy and how to deploy different versions of communication units for different countries. A vehicle has a life cycle of more than 15 years, and onboard communicating devices should not become obsolete during that period. Moreover, the drivers as well as passengers would like to use the newer technologies that come out in the future.

Cognitive radio combined with SDR offers a solution to cope with evolving and numerous technologies. Reconfigurable and reprogrammable capabilities of CR and SDR provide the possibilities to design 'future proof' onboard units that are upgradeable with software updates. This allows flexibility in deploying different versions of units for different countries and allows upgradeability when new technologies come out in future.

An OBU can integrate an SDR transceiver with navigational devices (GPS) and onboard computer to provide capabilities of context awareness and adaptation.

3.6 Highly mobile environment

Deploying base stations to provide wireless services (e.g., Internet access) in vehicular networks is challenging due to the highly mobile environment. VANETs are considered as a special case of mobile *ad hoc* networks (MANETs) due to their specific characteristics such as special mobility pattern, varying vehicle density, and interference with other types of networks.

Several applications of VANETs have been proposed that take into account the above-mentioned constraints of vehicular communication. As described before, different communication technologies have been standardized (WAVE, IEEE 802.11p, etc.) for vehicular communications. However, presently, the deployment and performance evaluation of the proposed standards, especially in a large-scale vehicular environment, are works in progress. Moreover, different standards and protocols proposed for vehicular communication applications should be interoperable. Additionally, short-range communication protocols may not be able to provide good Internet connectivity for high-speed vehicles and that would require developing new efficient protocols suitable for them.

4 RECENT ADVANCES IN CR-VANETS

Cognitive radio technology [36] presents a promising solution for addressing the problem of spectrum scarcity Vehicle communication may benefit from cognitive radio technologies such as dynamic spectrum access, adaptive software-defined radios, and cooperative communications as shown in [60–62]. CR-enabled VANETs can use additional spectrum opportunities in TV bands according to the QoS requests of the applications.

Existing works on CR [36, 58] focus on various issues that include investigating techniques for spectrum sensing and spectrum access (dynamic spectrum access), cooperative communications, MAC protocols [63], routing protocols, QoS [64], and software-defined radios. While several studies exist in literature on applying CR to wireless mesh networks, *ad hoc* networks, and cellular networks, the research on applying CR to VANETs is still in its early stage. The research solutions proposed for general-purpose CR networks cannot be directly applied to CR-VANETs as the unique features of vehicular environment, such as the role of mobility, and the cooperation opportunities need to be taken into account while designing the spectrum management functions for CR-VANETs. Thus, the existing solutions need to be customized for CR-VANETs to account for high mobility, dynamic topologies, frequent disconnections, etc. Unlike static CR scenarios, in CR-VANETs, multiple cooperating vehicles (during busy hours) can exchange spectrum information to get information on the spectrum availability. Moreover, this enables adaptive operations and proactive response feasible for the vehicles that follow.

In this section, we present some of the recent works that focus on using cognitive radio technology for vehicular networks. We provide a taxonomy of recent advances in CR-VANETs, in Figure 6. Different approaches are classified into spectrum sensing, spectrum management and QoS, and network. These approaches as well as their sub-categories related to cooperative sensing, sensing in fast varying channels centralized to distributed spectrum management approaches, routing, mobility management, content distribution, etc. are discussed in the following text.

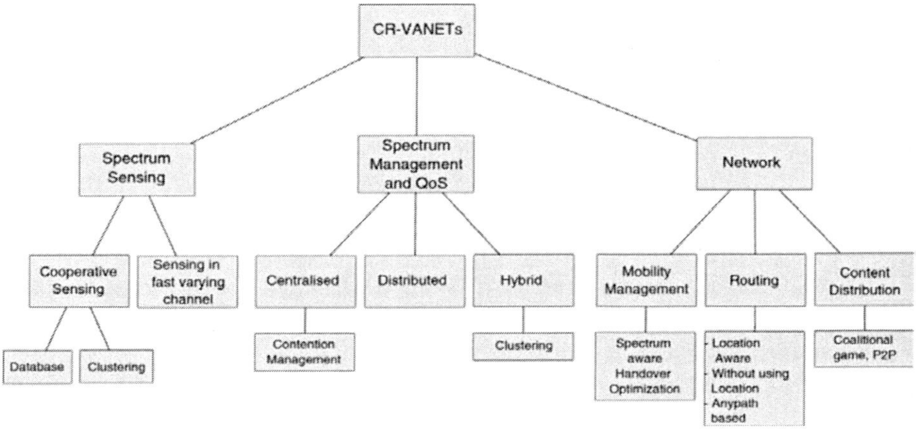

Figure 6 A TAXONOMY OF RECENT ADVANCES IN CR-VANETS.

4.1 Architecture

The network architecture of CR-VANETs consists of vehicles with onboard units (OBUs) and infrastructure units such as RSUs and base stations of network providers. Note that setting up of roadside units everywhere may be costly; thus, V2V communication, including multihop, should be able to work without any infrastructure support.

Several functionalities are needed for CR-VANETs paradigm, and an initial architecture related to different functionality blocks and protocols is proposed in [65]. The architecture proposed by its authors consists of several blocks related to policies such as those imposed by FCC and local authorities and onboard sensors, including radio sensors and GPS. It also includes a knowledge database that consists of previously learned cases, rules, spectrum information, and radio maps. Further, it consists of a cognitive engine that is the core of all functionalities. Additionally, there are software-defined radio or multiple radios, performance measurement block for feedback, and finally a user interface to allow users to configure the system offline. Figure 7 illustrates the architecture showing these different functional blocks for CR-VANETs.

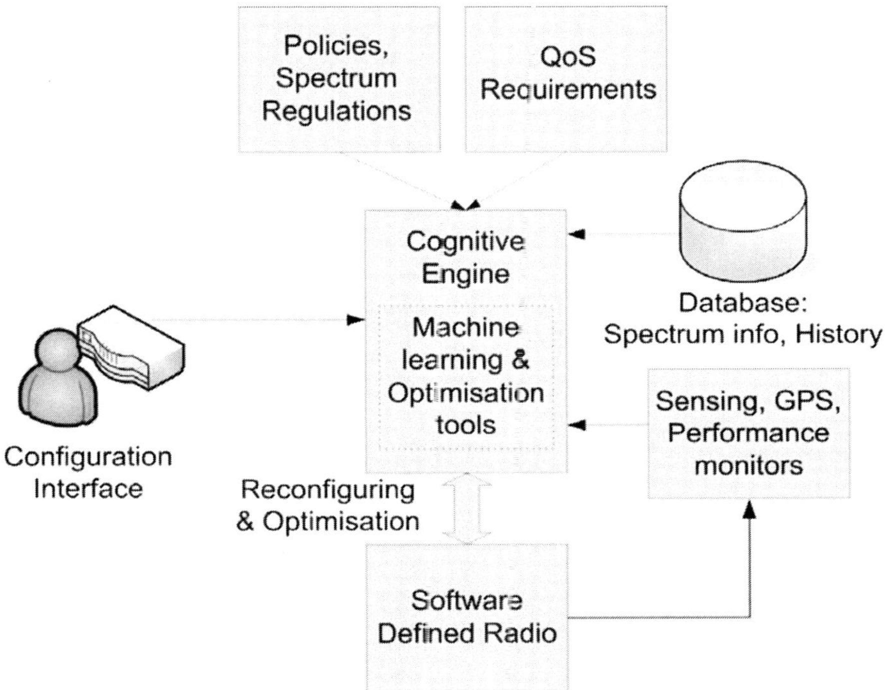

Figure 7 CR-VANETS FUNCTIONAL BLOCKS (ARCHITECTURE).

As illustrated in the figure, the cognitive engine (CE) uses real-time data from onboard sensors and history and map information from knowledge database to learn from the present as well as the past. It then controls the operation of software-defined radio through optimization of transmission parameters and decision making related to spectrum use and management.

A taxonomy of machine learning approaches for CE is provided in Figure 8. The classification is done in terms of supervised learning, decision making and parameter tuning. Supervised learning can be used for prediction or classification, by learning the mapping between the given inputs and outputs. It is called supervised learning because the outputs for the given inputs are known beforehand. For example, in the case of signal identification, the target signal and the input signal features are known during the learning phase. Decision making is related to the problem of making optimal decisions. Some available tools for decision making use reinforcement learning, which can be used in unknown environments, or case-based reasoning based on history, etc. Parameter tuning is needed to set the parameters optimally. A system model is used, and some metaheuristics can be used to find the optimal values of the system parameters for a given optimization objective. Multiple conflicting optimization objectives, such as maximization of bandwidth vs. minimization of bit error rate, can be assigned with weights differentially according to the desired mode of communication, ranging from bandwidth maximization mode to emergency mode for high resiliency [59], etc.

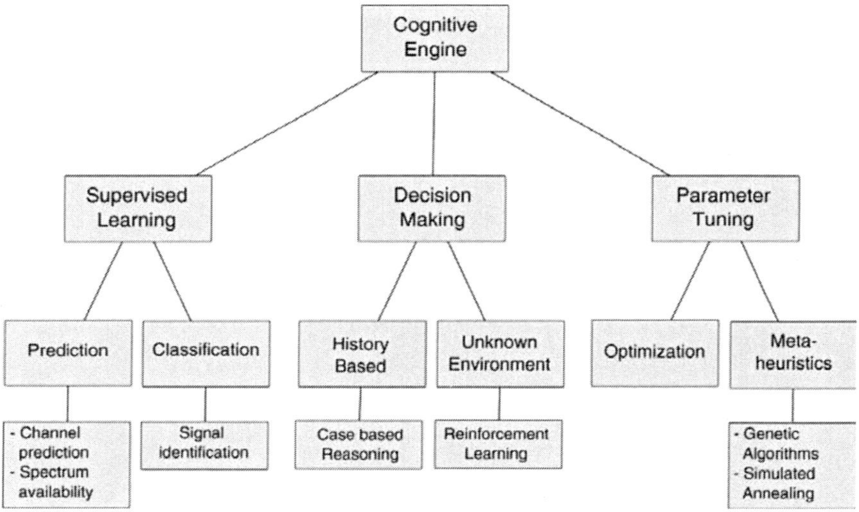

Figure 8 A TAXONOMY OF MACHINE LEARNING APPROACHES FOR COGNITIVE ENGINE.

Several machine learning techniques have been studied in the literature to realize cognitive functionalities in the general context of cognitive radio. However, they need to be adapted for CR-VANETs scenario. Some interesting ideas for applying machine learning to CR-VANETs are discussed in [61]. We summarize those ideas in Table 2 and discuss them in the following text.

Table 2 MACHINE LEARNING IN CR-VANETS (SUMMARIZED FROM [[61]])

Machine learning tools	Application
Case based reasoning	Faster convergence from exploiting previous solutions for example when traveling through same area on same day of the week
Support vector machine and neural networks	Signal classification and identifying spectrum holes, channel prediction
Reinforcement learning	Channel selection, real-time learning, can be used in unknown environments
Genetic algorithms and simulated annealing	Optimization of transceiver parameters

The authors of [61] argue that case-based reasoning is interesting for CR-VANETs scenario because vehicles have a high chance of passing through the same locations at approximately the same time of the day. This is especially the case for example, for public transports, such as buses and trains but also for cars considering that people commute everyday from their homes to offices, at approximately the same time. With case-based reasoning, the optimization process can be populated from the previously found optimal configurations, and then further search for optimization can start from there. This can highly improve the convergence time of optimization algorithms such as genetic algorithms and simulated annealing, which in turn can be used for optimizing transmission parameters.

Machine learning tools for classification such as support vector machines and neural networks are interesting for signal identification and discovery of spectrum holes. With cognitive radio, it is required that the opportunistic use of spectrum should not collide with primary user traffic. Thus, classification tools can be used to enhance the detection of primary user signals and can provide improved accuracy as compared to just using energy detection. This improved primary user detection accuracy is desired especially in VANETs scenarios, where high-mobility conditions leading to signal fading can cause difficulties in spectrum hole detection. Neural networks can also be used as function approximators to predict throughput at a given location on a particular time of the day. Different inputs such as

PU and SU traffic statistics can be mapped to predict throughput through trained neural networks [61]. Some online learning tools like reinforcement learning can be used. Such tools are useful in unknown environments and are useful for channel selection and decision making.

4.2 Spectrum sensing

Spectrum sensing is required by cognitive radio to detect the presence of spectrum holes and more importantly to detect the presence of PUs. The priority is to avoid interference with PUs and thus, spectrum sensing should be highly reliable.

High mobility in VANETs can deteriorate the performance of spectrum sensing due to signal fading and fast changing location. In the case of *fast varying channels* over time, some works focus on improving the performance of sensing [66] and signal identification [67–69]. Similarly improved algorithms have been proposed for channel estimation and equalization [70] in high-speed environments.

Additionally, *cooperative spectrum sensing*[71] can be used to tackle the problems due to high mobility. Several spectrum sensing vehicles combined with a regularly updated central spectrum database can improve the performance of spectrum sensing in VANETs scenarios. The work in [72] investigates the effect of mobility on the performance of spectrum sensing in CR networks. The study confirms that cooperation can improve the spectrum sensing performance as a result of increased spatial-temporal diversity in received primary signal strengths.

This forms the basis of their proposed sensing technique that finds an optimal combination of the number of sensors for cooperation and the number of times to sense the spectrum in order to reduce the sensing overhead. This work does not specifically target CR-VANETs scenario, but it is still interesting for that. Additional works such as [31, 34, 73] aim to achieve better spectrum sensing precision and efficiency through a coordinated spectrum sensing. The data sensed by the vehicles is sent to RSUs that forward the gathered data further to a central processing unit. The delay and reliability of spectrum sensing in vehicular environments is discussed in [74]. They argue that cooperative spectrum sensing can be used to tackle the sensing limitations of a single vehicle as it exploits spatial diversity. They show that sensing time can be reduced significantly with cooperative sensing as compared to sensing by only a single vehicle.

The authors of [75] point out that cooperative sensing can help a vehicle to detect the spectrum availability in future locations by exploiting sensing information from other cooperating vehicles. An experimental study on sensing performance in VANETs scenario is conducted in [30]. They found that sensing accuracy depends on vehicular speed as well as location. For example, in the areas with open space, the fast fading is not high and hence, sensing performance is good unlike at places such as downtown. They also propose a framework for cooperative spectrum sensing and sharing called Cognitive V2V (Cog-V2V). In Cog-V2V, each vehicle keeps a *spectrum availability database* that is regularly updated. Note that using such knowledge database, shared by different vehicles or kept distributed among them, is an interesting way of collaborating for spectrum sensing. The data from different cooperating vehicles is merged using weights that depend on distance between the locations at which sensing samples were collected. Intuitively, they calculate weights such that weights from nearby vehicles are weighed lower as they are likely to have similar radio conditions due to their similar location. After the spectrum opportunities are detected, Cog-V2V decides the channel to be used between the two vehicles.

Another work [76] uses spectrum database as well as support from infrastructure. They also provide guidelines for the uniform placement of RSUs that will provide information about the availability of TV channels and thereby support the mobile CR-enabled vehicles. Systems using centralized approaches and RSUs can be effective, specifically for safety applications as it improves their reliability by improving data delivery in vehicular networks.

However, the centralized approach is not apt to support multihop communication, especially in the absence of infrastructure support. Nonetheless, in the absence of central infrastructure, clustering techniques can be used to create cluster leaders among several vehicles. Cluster leaders can then assume the role of central entities responsible for coordinating spectrum sensing and data fusion as used in [35].

In [77], a different type of sensing is used that focuses on estimating the amount of congestion in a given channel. The authors address the problem of insufficient spectrum for reliable exchange of safety information over congested urban scenarios through the use of cooperative spectrum sensing provided by vehicles. The spectrum sensing provided by vehicles is used to provide a distributed measure of contention on the roads and locate white channels in the industrial, scientific, and medical (ISM) bands. If contention is detected, free channels from the ISM spectrum are allocated to the vehicles. Note that some of these works are also discussed in the next section in the context of spectrum management.

4.3 Spectrum management and QoS support

One of the main focuses of research in CR-VANETs is to design specific spectrum management techniques which take into account spectrum sensing and access as well as mobility for CR-enabled vehicles. Spectrum management should also aim at providing *QoS support* in CR-VANETs scenario. QoS guarantees are important for messages and signals related to vehicular safety. QoS guarantees can be in terms of maximum permitted delay and reliability in terms of permitted packet loss, etc. For example, in the event of an emergency brake, the vehicles that follow should be notified as quickly as possible in order to leave some time for other drivers to react.

However, providing such QoS guarantees can be difficult in case of congestion or when there are several vehicles. Moreover, it is noteworthy that the feasibility of using TV white space for vehicular communications is restricted due to the scarcity of TV white space in large urban cities, where the average number of vehicles is also higher as compared to the remote areas. Such factors can make it difficult to ensure QoS support for some important applications.

Taking the example of safety messages, authors in [77] argue that the bandwidth of a single DSRC control channel may not be enough for reliable delivery of safety messages during congestion. They propose to use cognitive radio to augment the spectrum of the control channel. They proposed to use TV white space (TVWS) spectrum to enhance the bandwidth of the control channel [15] and in a follow-up work [77], 5.8 GHz ISM band is used instead of TVWS, as 5.8 GHz ISM band is the neighboring band to the DSRC 5.9 GHz band and it can simplify the transceiver design. They argue that Wi-Fi users, who use 5.8 GHz ISM band, are unlikely to be present on highways.

In order to counter congestion problems, authors in [77] first propose a *contention metric* $C_r(t)$ to estimate the amount of congestion in different regions r, at time t:

$$C_r(t) = \alpha \frac{D_r \times B_r}{S_r} + \beta U_r. \quad (1)$$

where the delay D_r is the delay experienced by the transmitted packet and it increases during congestion. The parameter B_r is the channel's bitrate, S_r is the average packet size, and U_r is the number of untransmitted packets due to congestion. The parameters α and β are tuning parameters, which are estimated from curve fitting of simulation data. Different scenarios were simulated to obtain such data by varying the number of vehicles, payload sizes, and bitrates.

Different vehicles perform sensing and report the measurements to their nearest RSU. RSUs communicate with other RSUs and vehicles to form an estimate of congestion in different zones. Based on congestion estimates, a fuzzy-logic-based controller is used to decide the amount of spectrum needed, and RSUs allocate more spectrum to the vehicles traveling in a zone with increased congestion.

The idea is that the spectrum is a scarce resource and additional spectrum is allocated only when needed. Additionally, the proposed contention metric looks interesting and it will be interesting to see how it can be used in other 802.11-like scenarios. Another interesting research direction is to explore improved contention metrics.

Several interesting ideas related to dynamic spectrum access in CR-VANETs are explored in [44]. They call the dynamic spectrum allocation problem for VANETs as vehicular dynamic spectrum access (VDSA). Several VDSA approaches covering spectrum measurements and machine learning are proposed; for example, reinforcement learning is used for intelligent channel selection (also presented in [65]). Moreover, a testbed implementation with some experimental results is also provided. A *feasibility analysis* is done for VDSA using queuing theory. Different TVWS bands are considered as servers which are either occupied by PUs or not. Taking the case of V2I communication, the requests to access those channels from SU vehicles are modeled as clients arriving in a queue and these requests stay in the queue until they are served. Thus, with this formulation, the queuing theory results can be used to analyze VDSA scenario. However, it is pointed out in [78] that such model can be over optimistic because vehicles can move out of RSU range. Thus, vehicular dynamics need to be taken into account and an analysis considering such dynamics is proposed in [78].

Additional works such as in [30, 31] propose to first detect the available spectrum resources, then according to the QoS requirements of the vehicular applications, spectrum decision algorithms select the channel to be used at a specific location. Finally, note that even if vehicles have sufficient power supply, the *energy consumption* is still an important factor for environmental considerations [79]. An energy-efficient technique is proposed in [79] by formulating the problem of deciding whether to use direct transmission or relay-based transmission as an optimization problem, with the objective of minimizing energy subject to QoS constraints in terms of delay.

Most of the above approaches assume a centralized architecture where decisions can be taken in a central manner. However, this may not be always possible in the case of CR-VANETs. Thus, in [35] the authors focus on a *clustering strategy* as illustrated in Figure 9 where vehicles are grouped into clusters and one of the vehicles act as a cluster leader making central decisions for the cluster. It can be seen as a hybrid of distributed and centralized approaches. The authors study the problem of optimal channel access to provide QoS for data transmission in CR-VANETs. The data from cluster members is first transmitted to the cluster leader and then it is forwarded to its destination.

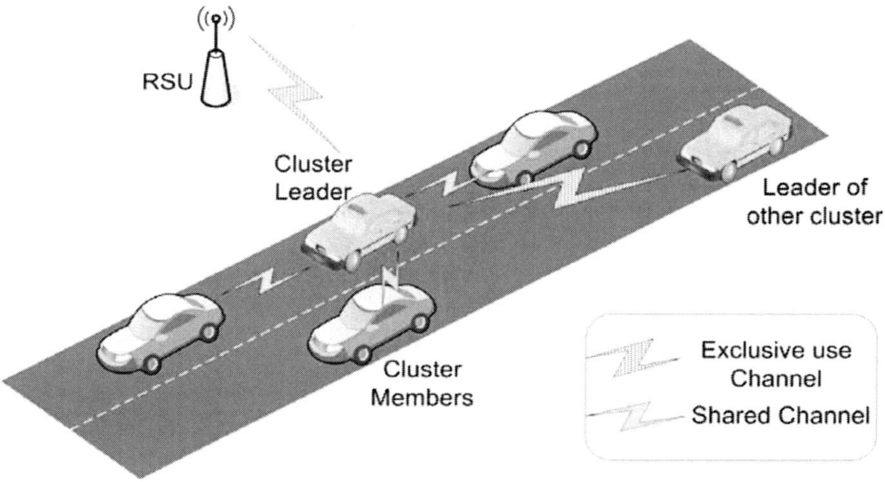

Figure 9 CLUSTERING STRATEGY. Vehicles are grouped into clusters and one of the vehicles act as a cluster leader making central decisions for the cluster.

The problem is to opportunistically use available channels using CR, optimally reserve some bandwidth in DSRC channels, and control the size of the vehicular cluster. The authors use constrained Markov decision process (MDP) for decision making. The vehicles use two radios: one for opportunistic spectrum access and the other for DSRC channel. Vehicles send a request to join a given cluster using DSRC channel. The cluster head takes the decision of whether the new node should be added to the cluster or not. The cluster head also optimally reserves some bandwidth in the DSRC channel. For opportunistic spectrum access, the cluster head collects the sensing results from cluster members and then the scheduling decision is broadcasted to the cluster. The authors point out that there are two different time scales involved in decision making: opportunistic channel access (fast) and cluster size control with DSRC bandwidth reservation (relatively slow). Thus, a hierarchical MDP with two levels with each level corresponding to each problem is used. These two levels interact with each other using the following parameters: QoS performance measures, location, cluster size, and available bandwidth in DSRC channel. The optimization problem is solved using linear programming, and different statistics and variables are assumed to be available in advance. Thus, one interesting extension of this work could be to explore online learning.

Some distributed spectrum management schemes exist for *ad hoc* CR networks that focus on MAC design [63] as well as QoS [64]. Such proposals may be customized for CR-VANETs. The distributed opportunistic multiChannel MAC (OMC-MAC) proposal in [64] assumes a fixed CCC, transmits beacons at regular intervals for synchronization, and divides the beacon interval into three phases: channel sensing, contention, and data transmission. SUs sense the channel in the first phase and then using CCC contend for reservations of time slots belonging to data transmission phase. Data transmission phase is then used for contention-free transmission of data based on the reservations on different available channels. QoS provisioning is done by first letting only the prioritized users contend for the reservations and if some time slots are left, then other users can also make some reservations. This can be useful for prioritizing safety messages in CR-VANETs. However, the impact of vehicular mobility and short-lived links needs to be studied for such schemes. The common control channel may also get congested when there are many vehicles. The discovery process of available channels, during the sensing phase, can be improved by using a common spectrum database, as explained in Section 4.2. Adaptive bitrate as well as beacon interval can be explored to further improve the performance.

4.4 Network layer: routing, mobility management, and content distribution

In this section, we discuss the works on CR-VANETs that are related to upper layers. We mainly focus on network layer but also talk about P2P approaches that work on top of an overlay network.

4.4.1 Routing

Several cognitive routing schemes have been proposed for CR-VANETs, accounting for spectrum scarcity, interference, and high mobility. Some important metrics for routing in CR-VANETs are PU activities, availability of spectrum, geographic location, channel switching delays, hop count, etc.

One initial protocol called Search [80] is proposed for cognitive radio *ad hoc* networks that is interesting for CR-VANETs scenario also. It selects a route and the channel that avoid the regions of PU activities. It sends route request packets (RREQs) and they reach the destination via different paths. The destination selects the optimal paths based on geographic forwarding. The cost of switching a channel is also taken into account because with cognitive radio the intermediate nodes might have to switch a channel to forward a packet on a given path. Then the paths are combined to minimize the number of hops. Another recent approach based on RREQ packets is proposed in [81] but without using the location information as all vehicles may not have GPS information. Two routing schemes are proposed: (1) First scheme searches through several routes with each route evolving on a single channel; (2) Second scheme discovers only one route, but it exploits several channels.

A recent protocol CoRoute [33] is an anypath vehicular routing protocol for CR-VANETs that exploits geo-location and sensed channel information. CoRoute extends the anypath multirate scheme proposed in [82] to include cognitive radio and geo-location. With *anypath multirate* routing, a node broadcasts to a set of nodes using a selected rate. Then, any one of them forwards the packet towards the destination in the same way while the other nodes in the set avoid broadcasting duplicate copies if they hear the first broadcast. This routing has an anypath property that any node that received the packet can forward it and hence, the path can change depending on which node received the packet. If multiple nodes receive broadcast then, the one with highest priority forwards the packet. This priority is chosen based on the expected anypath transmission time (EATT), which is the expected time to transmit a packet to the forwarding set depending on packet size, bitrate, and the delivery probability that the packet is received at least by a single node in the forwarding set.

CoRoute extends the anypath multirate routing protocol such that the broadcasts are done on different channels depending on the sensed channel information and geo-location. The set of forwarders is chosen such that the nodes in the set are geographically closer to the destination. The channel with the least amount of congestion is chosen. The priority of the nodes in the set of forwarders is set based on EATT as described before.

4.4.2 Mobility management

Mobility management is important for ensuring connectivity even when the vehicles move. The Internet Engineering Task Force (IETF) has defined some IP mobility support protocols that will be used for vehicular networks. For instance, NEMO [83] protocol based on Mobile IPv6 (MIPv6) [84] is being considered for vehicular scenarios. In addition, Proxy Mobile IPv6 (PMIPv6)-based NEMO protocol is also a possible candidate [85–87]. Currently, Distributed Mobility Management (DMM) and data offloading techniques are being considered [88]. DMM can reduce the overall path latency and can eliminate the single bottleneck problem of the centralized architectures. Moreover, data offloading is useful for adaptively offloading some flows to other available networks and benefiting from additional available bandwidth.

These mobility management approaches can benefit from the information available from cognitive radio modules. Network selection and handovers can be optimized, using cross-layer optimization, for example, by considering geo-location and spectrum availability to store different signal thresholds used for handover. The spectrum can be allocated among RSUs so as to maximize the coverage and minimize the number of handovers, etc. Additionally, some flows can be adaptively offloaded to the cognitive radio network using data offloading techniques.

Very few works [89] discuss mobility management in the context of cognitive radios. In [90], a joint mobility and spectrum management scheme is studied focusing on cellular networks, but the ideas are also interesting for CR-VANETs.

4.4.3 Content distribution using peer-to-peer techniques

For content distribution in CR-VANETs, one proposal [91] uses V2I communication for content distribution and uses V2V communication with cognitive radio to exchange the missing packets. They use a P2P-like content sharing approach using cognitive radio. For P2P communication in the context of CR-VANETs, the problem is to determine that the onboard unit should download which chunk, from which peer, depending on the link and spectrum characteristics. The authors propose an approach based on coalitional graph game. When some vehicles pass a RSU, then they can download some data chunks from RSU and other chunks from peer vehicles. For opportunistic spectrum use, each vehicle senses a set of spectrum channels for detecting the presence of primary users. A channel is chosen which has maximum transmission capacity and is not occupied by primary users or neighboring vehicles.

A utility function is used such that payoffs are a function of the number of successful packets transmitted as well as received multiplied by the probability of successful transmission. The probability of successful transmission is computed by the probability that no neighbor of a given vehicle is transmitting on a particular channel multiplied by the probability that the given transmission does not collide with primary user. The latter depends on the probability of missing the primary traffic and probability of false alarm. The cost of using a link between two vehicles is defined as a weighted function of the total number of neighbors of receiving and transmitting vehicles. This cost function is used to represent the probability of data collision because any transmission on this link has a chance of interfering with those neighbors.

Finally, a myopic distributed algorithm is proposed whose objective is to converge towards a graph describing which vehicle connects with which other vehicle or RSU. The proposed algorithm is myopic as it uses only local information, as compared to centralized algorithms which need global information. Initial approach is that each vehicle chooses a graph that just optimizes its own utility based on current neighbors and channel characteristics.

Usually, it is desired for any distributed algorithm to converge. However, authors state that it is difficult to prove the convergence of such algorithms due to dynamically evolving topologies in VANETs scenarios. Moreover, the proposed algorithm can have a ping pong effect that is to oscillate from one graph formation strategy to another. Thus, they introduce a notion of history function which keeps a count of how many times a given strategy was used. If a strategy was already used up to a certain threshold, then it is not used again. With this history function, the authors show that their algorithm converges within a finite number of iterations. The simulation results show that their algorithm can improve the throughput by 250% as compared to just V2I broadcast.

It would be interesting to see how this work can be extended for the case of multihop communication (only line of site neighbors are used) and selfish users. Additional ways of ensuring convergence could be explored with newer algorithms. Moreover, it would be interesting to consider more detailed link models as presently only a simple model for link layer collisions and contention is used.

5 PROJECTS AND TESTBEDS

The research on cognitive-radio-enabled vehicular networks is still in its preliminary stage, and there are not many related experimental platforms due to their complex setup and requirements. In this section, we describe some existing projects and testbeds related to cognitive-radio-enabled vehicular networks.

5.1 CORRIDOR

Cognitive Radio for Railway Through Dynamic and Opportunistic Spectrum Reuse (CORRIDOR) [92] is a French research project that targets opportunistic spectrum access for railways. Communication demands are increasing for modern railways from the point of view of railway operations as well as for providing Internet connectivity to the passengers. However, there is no single universal wireless technology that can answer the needs of heterogeneous services and railway applications. Modern railway applications and passenger's Internet connectivity demands call for more bandwidth and spectrum.

One answer to such needs is cognitive radio. Thus, CORRIDOR plans to use CR technology to support multiple railway applications such as command and control, CCTV, maintenance, and Internet connectivity for passengers. The project aims to develop algorithms and technology for very high speed environments that are typical of railways. The focus will be on spectrum access and management, exploiting TVWS, cognitive engine, cross-layer optimization, mobility management, and handover optimization. Different techniques will be validated with some real on-site trials.

5.2 PROTON-PLATA project

The PROTON-PLATA project [93] is a European project that aims at developing a reconfigurable prototype based on emerging software-defined technologies for telematics applications for cars-to-roadside and car-to-car communications. The project proposes a multitechnology cooperative Advanced Driver Assistance System (ADAS) that is based on the integration of SDR devices in vehicles.

The first part of the project deals with developing a prototype including SDR device and driver assistance applications to demonstrate the utility of a cooperative ADAS for improving communication performance. This involves designing a multitechnology communication infrastructure and equipping onboard units with SDR devices that support different communication scenarios i.e., V2V communications, V2I communications, and broadcasting (infrastructure-to-vehicle) at the same time. The performance evaluation using network simulation tools forms the second part of the project.

5.3 Rail-CR project

The Rail-CR project [94] is a US project that is a part of the efforts toward implementing the positive train control (PTC) technology to equip trains with wireless communication capabilities. This will allow trains to communicate with wayside wireless stations, while moving, and provide useful information related to their location, speed, direction, etc. The project adapts Virginia Tech cognitive radio technology [95] to improve safety and operations of the railways.

The complex radio characteristics faced during train transportation such as continuously changing situation of the train as well as continuously changing communications environment due to varying noise, multiple sources of interference, multiple users competing for scarce spectrum, etc. make wireless communication with trains very demanding. This has motivated the railway industry to use SDRs equipped with a reconfigurable platform for packet-data transmission. Though SDR provides interoperability and enables multiple configurations, it lacks the capability of integrating additional adaptation to alleviate crowded spectrum, intentional jamming, or learning from past experiences.

In this context, the project developed a railroad-specific cognitive radio namely, *Rail-CR*, capable of fulfilling the requirements of future wireless communication systems for trains. The Rail-CR combines AI-based decision-making and learning algorithms with SDR to address the railway transportations' specific need for adaptable communications. The Rail-CR cognitive radio will enhance the performance of railway communications in terms of interoperability, robustness, reliability, and spectral efficiency, and will make it cost-effective to deploy and maintain.

The Rail-CR system includes a CE that relies on the tunable radio parameters (knobs) and observable performance indicators (meters) of the SDR platform. Based on situational awareness from the radio in the form of observable parameters (meters), a cognitive engine uses software-based decision-making and learning algorithms to decide if the radio parameters (knobs) should be adapted depending on sets of predefined objectives. The Rail-CR was evaluated under different interference conditions, and results show that the cognitive engine can manage to overcome the interference by adapting configurable parameters whereas a radio minus cognition capability was not able to overcome interference resulting in high errors or a loss of connectivity.

Further, a case study showing the use of metacognitive radio to improve wireless communication systems used for signaling and train control in railroads is presented in [96]. In a metacognitive engine [96], a master process monitors and adapts the cognition process (i.e., cognitive engine). Metacognition enhances the performance of a cognitive radio with better learning and decision-making capabilities.

5.4 The cognitive cars testbed

Research in VANETs has focused on simulation studies to evaluate the performance of communication protocols for VANETs. Testing and evaluation of VANETs communication protocols (e.g., VANETs-based multihop communication system) in real testbed scenarios for example, on the road, has been considered a difficult task as there are virtually no highly dense vehicular testbed that exists. Recently, a new cognitive approach to VANETs research was introduced in [97] that enables performing real vehicular experiments, which usually require several vehicles, using only a few vehicles equipped with wireless communication interfaces. This is done by setting up a virtual overlay network consisting of relaying and interfering vehicles on top of a group of only a few vehicles to conduct experiments. The communication packets travel over a random number of hops and experience interference from random number of vehicle transmitters with varying transmission characteristics.

The cognitive car testbed [98] is based on guidelines presented in [97] to conduct real experiments with VANET applications and communication protocols under constraints of vehicular environment and computing resources. The objective of the cognitive car testbed is to explore the possibility of using cognitive network technologies in advanced vehicular testbeds. It focuses on using cognitive radios to study the impact of different frequency bands and different frequency switching delays on the performance of VANET communication protocols. More specifically, cognitive car testbed enables evaluating a communication protocol vs. different variables including number of hops, density of interfering neighbors, wireless channel conditions, transmission frequency, and switching delay at the node.

In this context, it introduces the architecture guidelines for implementing a VANET testbed and, in addition, provides preliminary results for the experiments conducted with an accident warning system developed for highways and a cognitive network based on the software radio (SORA) technology provided by Microsoft. Preliminary results show that cognitive interfaces can be used as an additional tunable dimension in an experiment platform where highly dense vehicular testbeds can be set up with just few real vehicular resources (e.g., few cars and drivers). The cognitive interfaces enable testing novel techniques designed for addressing radio spectrum scarcity in a vehicular environment as well as performance of VANET communication protocols. The performance of VANET communication protocol can be evaluated by tuning two additional variables: frequency band and switching time. The first variable is the frequency band used for transmitting and receiving communication packets and the second variable is the switching time needed to change from one frequency to another.

5.5 Virginia Tech CR network testbed (VT-CORNET)

VT-CORNET [95] is an open-source cognitive radio network testbed, developed by Virginia Tech (Blacksburg, VA, USA) for the evaluation of CR wireless communication protocols and applications. The testbed consists of 48 software-defined radio nodes (USRP2-based nodes) deployed over the four floors of a building and equipped with a custom-made daughterboard covering the frequency range of 100 MHz to 4 GHz. The CORNET nodes are remotely accessible to the registered users. A unique feature of the testbed is that it uses Software Communications Architecture (SCA) framework that assumes USRP nodes at the physical layer and provides support for generating and visualizing a range of radio configurations.

The testbed focuses on CE design, self-organizing networks, and topology research. VT-CORNET is a highly reconfigurable testbed that enables testing of independently developed CR engines, sensing techniques, applications, protocols, performance metrics, and algorithms. In addition to the static 48 nodes deployed in the ceiling, low-power mobile nodes will also be available to support vehicular environment. The rail-CR project has adapted the VR-CORNET cognitive radio technology for railways as described in the first part of this section.

The CORNET testbed is openly available for conducting research on advanced CR networks. It provides a collection of resources to researchers lacking the ability to perform advanced experiments because of limited exposure to software-defined radio and CR platforms. The resources are made available to the researchers through flexible RF front ends, open-source software, and FCC experimental licenses.

5.6 ORBIT testbed

The ORBIT testbed [99] is an open-access research testbed for next generation wireless networking that allows large-scale reproducible experimentation on next generation protocols and applications. It consists of an indoor radio grid emulator for large-scale reproducible experiments and an outdoor field trial system for supporting real-world experimentation and evaluations.

A 400-node radio grid emulator has been deployed at the WINLAB Tech Center building and is open for general use by the research community to conduct experiments. The ORBIT testbed provides support for conducting a wide range of experiments including mobile *ad hoc* networks, dynamic spectrum coordination, network virtualization, wireless security, and vehicular networking. The radio grid testbed enables emulation of real-world network topologies via noise injection and packet filtering. The testbed has been recently upgraded to include programmable software-defined radios (GNU USRP/USRP2) for flexible MAC/PHY experiments.

The testbed users can access the radio grid through a web portal to specify their experiments and measurements. As mentioned previously, the testbed can be accessed worldwide for conducting experiments and evaluate protocols and applications related to *ad hoc* routing, P2P, spectrum sensing, and dynamic spectrum access. Recently, the testbed has started providing support for vehicular mobility as well. The ORBIT management software has been extended to enable mobile vehicular (vehicle-to-vehicle networks) experiments, and the radio grid is now supplemented by a number of outdoor and vehicular nodes.

6 OPEN RESEARCH ISSUES

CR-VANETs is a relatively new research area and hence, there are many issues that are open to further research. In general, customizing existing CR approaches to CR-VANETs is an interesting research direction. Additionally, there are some specific issues and we discuss some of them in this section.

6.1 Vehicular mobility

Initial studies [30, 72] on the effect of vehicular mobility on spectrum management show that the primary users may have to face adverse interference due to the incorrect detection of occupied frequencies as a result of the Doppler spread generated by vehicular mobility. At the same time, since a CR-enabled vehicle can gather samples of signal at different locations along its path while moving, this increases the spatio-temporal diversity of the samples and, in turn, decreases the chances of inaccurate decisions due to shadowing effects.

Open issues on how the performance of spectrum sensing is affected by vehicular mobility that need investigation include the following: techniques to achieve a balance between cooperation and spectrum scheduling for a CR-enabled vehicle on the move, how to use predictable mobility of vehicles to improve spectrum awareness, and how the mobility-related parameters such as speed and direction affect the performance of spectrum sensing [28].

6.2 Vehicle-to-vehicle communications

Mobile *ad hoc* networking is difficult in general. In case of CR-VANETs, V2V communication is more complex and difficult because of dynamically evolving topologies and frequent disconnections that are due to vehicles going out of range. This happens more often when the vehicular density is low. Such disconnections can create instability and performance issues for upper layer protocols such as routing and transport (TCP). Some cross-layer strategies to optimize transport protocols, for example TCP, can also be explored such as in [100, 101]. For routing in vehicular networks, some approaches based on delay-tolerant networking (DTN) are becoming popular. DTN approaches can help in improving the stability and performance of routing protocols by providing some tolerance to disruptions. Thus, CR-VANETs routing protocols also need to integrate some DTN elements to counter frequent disconnections, apart from being spectrum aware.

6.3 Machine learning and environment characterization in CR-VANETs

Several machine learning approaches have been applied for cognitive radio. However, as discussed in this paper, CR-VANETs can have more stringent requirements in terms of requiring fast algorithms due to fast changing environment. Moreover, there can be different types of redundancies that can be exploited such as redundancy in network statistics at a given location at the same time of a day. This is especially pertinent in the case of public transport such as buses and trains. Thus, machine learning algorithms that are more specific to CR-VANETs should be explored. Moreover, it is also important to design the knowledge database, which is consultable by machine learning algorithms for faster convergence. Some design issues are the following: which statistics to measure and store, how to divide the region into different zones, how to characterize different environments from the point of view of radio propagation (open space vs. city landscape), how to fuse different measurements, etc.

6.4 Distributed algorithms

Many of the works assume centralized decision making for spectrum access and sharing. This may be valid in cases with infrastructure support for example, when a RSU can take central decisions. However, setting up of infrastructure is difficult and using the services of cellular networks has some costs as well as bandwidth limitations. Thus, infrastructure support may not be available all the time and consequently, distributed algorithms are highly desired for CR-VANETs. Spectrum-aware distributed algorithms for CR-VANETs should be explored that only use local information, such as that available from neighboring vehicles. These algorithms should have low convergence time, incur low signaling overhead, and should be fault tolerant.

6.5 Adaptive beamforming

Directional antennas are an interesting way to reuse space through beamforming, by focusing the antenna pattern on a specific region. A recent work on analyzing multicast capacity with beamforming is [102]. Beamforming also improves throughput as signal quality is improved in the concentrated beam. Moreover, beamforming is more relevant in vehicles than as compared to, for example, smart phones. This is because smart phones rapidly change their orientation in our hands, which makes it difficult to exploit beamforming with smart phones. On the other hand, vehicles change their orientation only while changing the route and it can be used to enhance the communication. With adaptive beamforming, a vehicle can focus the antenna beam only on a particular vehicle and can thus avoid interference in another zones containing PUs or other SUs. Thus, adaptive beamforming combined with opportunistic spectrum access is an interesting direction to explore.

6.6 QoE aware optimization

Most of the works focus on optimizing QoS metrics, such as throughput and delay. Some works optimize user utilities based on generic utility functions. However, it is ultimately the quality of experience (QoE) which matters to the users. For example, perceptual quality, an important QoE component for videos, is important for passengers streaming a video. For video streaming applications, it is now easier to adapt to changing network characteristics with some new standards such as scalable video coding (SVC) and dynamic HTTP streaming (DASH). One of the optimization approaches, in wireless networks, can be to use subjective testing to design perceptual video quality models for SVC and DASH [103, 104] and apply such models for optimizing QoE, as done for wireless multicast in [105]. Thus, similar QoE aware approaches can be explored in the context of CR-VANETs by considering additional parameters related to spectrum characteristics and availabilities.

6.7 CR-VANETs experiment and simulation platforms

Typically, the hardware experiment platforms designed for VANETs and CR-VANETs consist of regular vehicles and software-defined radios [106, 107] and it is complex to set up such platforms involving vehicle mechanics integrated with radios. Thus, currently, simulators are used. Using traffic simulators such as SUMO [108] and network simulators (NS-2/3 and OMNeT++) together enable designing simulators that combine vehicle mobility and communication networks. However, there is a lack of suitable simulators with features such as spectrum database, spectrum sensing and management, power spectral density, wireless interference level, and Doppler effects in order to be able to model and simulate varying topology of CR-VANETs and local wireless networks.

6.8 Security and privacy

With the use of cognitive radio for vehicular networks, there are additional security-related issues which need to be addressed, such as protecting the privacy of cooperating vehicles and identifying malicious CRenabled vehicles that broadcast false spectrum sensing reports while on the move. Addressing security issues in CR-VANETs is challenging as neighboring vehicles can keep changing over time while moving. While moving on the road, it is difficult to detect a malicious car that may be transmitting fake spectrum sensing reports as this requires rapid detection and correction.

Moreover, it is important to consider the privacy concerns of end users while using collaborative spectrum sensing in CR-enabled vehicles. Security issues should focus on how to prevent a potential attacker from tracking a driver's identity and movements by eavesdropping the periodic spectrum information broadcast by the vehicle [28].

7 CONCLUSIONS

This paper has surveyed novel approaches and discussed research challenges related to the use of cognitive radio technology for VANETs. Though CR deployment in vehicular networks is still in the preliminary stage, CR-VANETs has the potential of becoming a killer CR application in the future due to a huge consumer market for vehicular communications. However, the research solutions proposed for general-purpose CR networks cannot be directly applied to CR-VANETs due to their unique features that need to be considered while designing the spectrum management functions for CR-VANETs. In this context, several challenges and requirements for CR-VANETs have been identified. We have provided recent advances and open research directions on applying cognitive radio for vehicular networks focusing on architecture, machine learning, cooperation, reprogrammability, and spectrum management as well as QoE optimization for infotainment applications. A taxonomy of recent advances in CR-VANETs is also provided. In addition, some existing testbeds and research projects related to CR-VANETs have been described.

DECLARATIONS

Acknowledgements

This work was supported by the CORRIDOR project (ANR-11-VPTT-004) of the French national research organization (ANR). We also thank the anonymous reviewers and Remi Painchault from Thales, for their valuable comments which helped in improving this article.

Competing interests

The authors declare that they have no competing interests.

REFERENCES

1. Rawat P, Deep Singh K, Chaouchi H, Bonnin J-M: Wireless sensor networks: a survey on recent developments and potential synergies. J. Supercomput 2013, 1-48.

2. Toor Y, Muhlethaler P, Laouiti A: Vehicle ad hoc networks: applications and related technical issues. IEEE Commun. Surv. Tutor 2008, 10(3):74-88.

3. Hartenstein H, Laberteaux KP: A tutorial survey on vehicular ad hoc networks. IEEE Commun. Mag 2008, 46(6):164-171.

4. Boukerche A, Oliveira HA, Nakamura EF, Loureiro AA: Vehicular ad hoc networks: a new challenge for localization-based systems. Comput. Commun 2008, 31(12):2838-2849. 10.1016/j.comcom.2007.12.004

5. Li F, Wang Y: Routing in vehicular ad hoc networks: a survey. IEEE Veh. Tech. Mag 2007, 2(2):12-22.

6. Luo J, Hubaux J-P: A survey of research in inter-vehicle communications. In Embedded Security in Cars. Berlin Heidelberg: Springer; 2006:111-122.

7. Cseh C: Architecture of the dedicated short-range communications (DSRC) protocol. In Vehicular Technology Conference, 1998. VTC 98. 48th IEEE, vol. 3. 1998, Ottawa: IEEE; 18–21 May 1998:2095-2099.

8. Zhu J, Roy S: MAC for dedicated short range communications in intelligent transport system. IEEE Commun. Mag 2003, 41(12):60-67. 10.1109/MCOM.2003.1252800

9. Cheng L, Henty BE, Stancil DD, Bai F, Mudalige P: Mobile vehicle-to-vehicle narrow-band channel measurement and characterization of the 5.9 GHz dedicated short range communication (DSRC) frequency band. IEEE J. Select. Areas Commun 2007, 25(8):1501-1516.

10. Kenney JB: Dedicated short-range communications (DSRC) standards in the United States. Proc. IEEE 2011, 99(7):1162-1182.

11. Amendment 6: Wireless Access in Vehicular Environments (WAVE), Part 11, IEEE Standard 802.11p, IEEE Standard 802.11p

12. Jiang D, Delgrossi L: IEEE 802.11 p: Towards an international standard for wireless access in vehicular environments. In IEEE Vehicular Technology Conference 2008. VTC Spring 2008. Singapore; 11–14 May 2008:2036-2040.

13. Regan MA, Oxley J, Godley S, Tingvall C: Intelligent transport systems: safety and human factors issues, no. 01/01 (Royal

Automobile Club of Victoria (RACV) Ltd., 2001),. Monash University, Australia, 2001

14. ETSI-Intelligent Transport. Intelligent Transport Systems . Accessed on 11 March 2014 http://www.etsi.org/technologies-clusters/technologies/intelligent-transport

15. Ghandour AJ, Fawaz K, Artail H: Data delivery guarantees in congested vehicular ad hoc networks using cognitive networks. In 2011 7th International IEEE Wireless Communications and Mobile Computing Conference (IWCMC). Istanbul; 4–8 July 2011:871-876.

16. Federal Communications Commission: Spectrum Policy Task Report of the Spectrum Efficiency Working Group. 2002.

17. McHenry MA: NSF spectrum occupancy measurements project summary. Shared spectrum company report. . http://www.sharedspectrum.com/

18. Federal Communications Commission: Spectrum Policy Task Force report. ET Docket (02-135) 2002, 215.

19. Zhao Q, Sadler BM: A survey of dynamic spectrum access. IEEE Signal Processing Mag 2007, 24(3):79-89.

20. Patil K, Prasad R, Skouby K: A survey of worldwide spectrum occupancy measurement campaigns for cognitive radio. In IEEE 2011 International Conference on Devices and Communications (ICDeCom). Mesra; 24–25 February 2011:1-5.

21. Federal Communications Commission: Second memorandum opinion and order in the matter of unlicensed operation in the TV broadcast bands (ET Docket No. 04-186) and additional spectrum for unlicensed devices below 900 MHz and in the 3 GHz band (ET Docket No. 02-380) (FCC 10-174). Washington, D.C.; September 2010.

22. Mitola III J, Maguire Jr GQ: Cognitive radio: making software radios more personal. IEEE Pers. Commun 1999, 6(4):13-18. 10.1109/98.788210

23. Mitola III J: Cognitive radio for flexible mobile multimedia communications. In IEEE International Workshop on Mobile Multimedia Communications, 1999. (MoMuC'99). San Diego; 15–17 November 1999:3-10.

24. Mitola III J: Cognitive radio: an integrated agent architecture for software defined radio. Dissertation. PhD thesis, Royal Inst. Technol. (KTH), Stockholm, Sweden, pp. 271–350 (2000)

25. Haykin S: Cognitive radio: brain-empowered wireless communications. IEEE J. Select. Areas Commun 2005, 23(2):201-220.

26. Akyildiz IF, Lee W-Y, Vuran MC, Mohanty S: Next generation/dynamic spectrum access/cognitive radio wireless networks: a survey. Comput. Netw 2006, 50(13):2127-2159. 10.1016/j.comnet.2006.05.001

27. Wyglinski AM, Nekovee M, Hou T: Cognitive Radio Communications and Networks: Principles and Practice. Academic; 2009.

28. Di Felice M, Doost-Mohammady R, Chowdhury KR, Bononi L: Smart radios for smart vehicles: cognitive vehicular networks. IEEE Veh. Tech. Mag 2012, 7(2):26-33.

29. Pagadarai S, Wyglinski A, Vuyyuru R: Characterization of vacant UHF TV channels for vehicular dynamic spectrum access. In 2009 IEEE Vehicular Networking Conference (VNC). Tokyo; 28–30 October 2009:1-8.

30. Di Felice M, Chowdhury KR, Bononi L: Cooperative spectrum management in cognitive vehicular ad hoc networks. In 2011 IEEE Vehicular Networking Conference (VNC). Amsterdam; November 2011:47-54, 14–16.

31. Li H, Irick DK: Collaborative spectrum sensing in cognitive radio vehicular ad hoc networks: belief propagation on highway. In 2010 IEEE 71st Vehicular Technology Conference (VTC 2010-Spring). Taipei; 16–19 May 2010:1-5.

32. Tsukamoto K, Omori Y, Altintas O, Tsuru M, Oie Y: On spatially-aware channel selection in dynamic spectrum access multi-hop inter-vehicle communications. In 2009 IEEE 70th Vehicular Technology Conference Fall (VTC 2009-Fall). Anchorage; 20–23 September 2009:1-7.

33. Kim W, Gerla M, Oh SY, Lee K, Kassler A: CoRoute: a new cognitive anypath vehicular routing protocol. Wireless Comm. Mobile Comput 2011, 11(12):1588-1602. 10.1002/wcm.1231

34. Wang XY, Ho P-H: IEEE Trans. Veh. Technol. 2010, 59(4):1936-1948.

35. Niyato D, Hossain E, Wang P: Optimal channel access management with QoS support for cognitive vehicular networks. IEEE Trans. Mobile Comput 2011, 10(4):573-591.

36. Akyildiz IF, Lee W-Y, Chowdhury KR: CRAHNs: Cognitive radio ad hoc networks. Ad Hoc Networks 2009, 7(5):810-836. 10.1016/j.adhoc.2009.01.001

37. ISO Technical Committee 204 WG16: Intelligent transport systems - communications access for land mobiles (CALM) - architecture. 2010.

38. ISO Technical Committee 204 WG16: Intelligent transport systems - communications access for land mobiles (CALM) - M5,. ISO 21215

39. ISO Technical Committee 204 WG16: Intelligent transport systems - communication access for land mobiles (CALM) - non-IP networking - Part 1: fast networking and transport layer protocol (FNTP),. ISO 29281-1:2013

40. ISO Technical Committee 204 WG16: Intelligent transport systems - communications access for land mobiles (CALM) - ITS station management - Part 5: fast service advertisement protocol (FSAP),. ISO 24102-5:2013

41. ETSI ES 202 663 v1.1.0.: Intelligent Transport Systems (ITS); European profile standard for the physical and medium access control layer of Intelligent Transport Systems operating in the 5 GHz frequency band.

42. Sevlian R, Chun C, Tan I, Bahai A, Laberteaux K: Channel characterization for 700 MHz DSRC vehicular communication. J. Electr. Comput. Eng 2010, 2010: 40.

43. Jakubiak J, Koucheryavy Y: State of the art and research challenges for VANETs. In 5th IEEE Consumer Communications and Networking Conference, 2008. CCNC 2008. Las Vegas; 10–12 January 2008:912-916.

44. Chen S: Vehicular dynamic spectrum access: using cognitive radio for automobile networks, Ph.D. Thesis, Worcester Polytechnic Institute. 2012.

45. Chen ZD, Kung H, Vlah D: Ad hoc relay wireless networks over moving vehicles on highways. In Proceedings of the 2nd ACM International Symposium on Mobile Ad Hoc Networking & Computing, Long Beach. New York: ACM; 2001:247-250.

46. Niculescu D, Nath B: Trajectory based forwarding and its applications. In Proceedings of the 9th Annual International Conference on Mobile Computing and Networking, San Diego. New York: ACM; 2003:260-272.

47. Bose P, Morin P, Stojmenović I, Urrutia J: Routing with guaranteed delivery in ad hoc wireless networks.Wireless Networks. 2001, 7(6):609-616.

48. Lee J-H, Ernst T: IPv6 security issues in cooperative intelligent transportation systems. Comput. J 2013, 56(10):1189-1197. 10.1093/comjnl/bxs006

49. Lee J-H, Ernst T: Security issues of IPv6 communications in cooperative intelligent transportation systems (poster). In 2011 IEEE Vehicular Networking Conference (VNC). Amsterdam; 14–16 November 2011:284-290.

50. Lee J-H, Ernst T, Bonnin J: Cross-layered architecture for securing IPv6 ITS communication: example of pseudonym change. In IEEE 2011 Third International Workshop on Cross Layer Design (IWCLD). Rennes; November 2011:1-6.

51. Lee J-H, Ernst T, Ma X: Performance analysis of secure beaconing messages for GeoNetworking. Secur. Comm. Netw 2011. doi: 10.1002/sec.396

52. Mitola J: The software radio architecture. IEEE Commun. Mag 1995, 33(5):26-38. 10.1109/35.393001

53. Mitola J: Software Radio. New York: Wiley; 2000.

54. Tuttlebee WH: Software Defined Radio: Origins, Drivers, and International Perspectives. New York: Wiley; 2002.

55. Dillinger M, Madani K, Alonistioti N: Software Defined Radio: Architectures, Systems and Functions. New York: Wiley; 2005.

56. Jondral FK: Software-defined radio: basics and evolution to cognitive radio. EURASIP J. Wirel. Commun. Netw 2005, 2005(3):275-283.

57. Arslan H: Cognitive Radio, Software Defined Radio, and Adaptive Wireless Systems. London: Springer; 2007.

58. Wang B, Liu KR: Advances in cognitive radio networks: a survey. IEEE J. Sel. Top. Sign. Proces 2011, 5(1):5-23.

59. Newman TR, Rajbanshi R, Wyglinski AM, Evans JB, Minden GJ: Mobile Network Appl. 2008, 13(5):442-451. 10.1007/s11036-008-0079-8

60. Chen S, Wyglinski AM, Pagadarai S, Vuyyuru R, Altintas O: Feasibility analysis of vehicular dynamic spectrum access via queueing theory model. IEEE Commun. Mag 2011, 49(11):156-163.

61. Chen S, Vuyyuru R, Altintas O, Wyglinski AM: Learning in vehicular dynamic spectrum access networks: opportunities and challenges. In IEEE International Symposium on Intelligent Signal Processing and Communications Systems (ISPACS). Chiang Mai; 7–9 December 2011:1-6.

62. Chen S, Wyglinski AM: Efficient spectrum utilization via cross-layer optimization in distributed cognitive radio networks. Comput. Comm 2009, 32(18):1931-1943. 10.1016/j.comcom.2009.05.016

63. Jha SC, Rashid MM, Bhargava VK, Despins C: Medium access control in distributed cognitive radio networks. Wireless Commun 2011, 18(4):41-51.

64. Jha SC, Phuyal U, Rashid MM, Bhargava VK: Design of OMC-MAC: an opportunistic multi-channel MAC with QoS provisioning for distributed cognitive radio networks. IEEE Trans. Wireless Commun 2011, 10(10):3414-3425.

65. Chen S, Vuyyuru R, Altintas O, Wyglinski AM: On optimizing vehicular dynamic spectrum access networks: automation and learning in mobile wireless environments. In 2011 IEEE Vehicular Networking Conference (VNC). Amsterdam; 14–16 November 2011:39-46.

66. Hassan K, Gautier R, Dayoub I, Radoi E, Berbineau M: Non-parametric multiple-antenna blind spectrum sensing by predicted eigenvalue threshold. 2012 IEEE International Conference on Communications (ICC) 2012, 1634-1629.

67. Kharbech S, Dayoub I, Simon E, Zwingelstein-Colin M: Blind digital modulation detector for MIMO systems over high-speed railway channels. In Communication Technologies for Vehicles. Heidelberg: Springer; 2013:232-241.

68. Hassan K, Dayoub I, Hamouda W, Berbineau M: Automatic modulation recognition using wavelet transform and neural networks in wireless systems. EURASIP J. Adv. Sig. Pr. 2010. doi:10.1155/2010/532898

69. Hassan K, Dayoub I, Hamouda W, Nzeza CN, Berbineau M: Blind digital modulation identification for spatially-correlated MIMO systems. IEEE Trans. Wireless Commun 2012, 11(2):683-693.

70. Simon E, Khalighi M: Iterative Soft-Kalman channel estimation for fast time-varying MIMO-OFDM channels. IEEE Wireless Commun. Lett 2013, 2(6):599-602.

71. Akyildiz IF, Lo BF, Balakrishnan R: Cooperative spectrum sensing in cognitive radio networks: a survey. Phys. Commun 2011, 4(1):40-62. 10.1016/j.phycom.2010.12.003

72. Min AW, Shin KG: Impact of mobility on spectrum sensing in cognitive radio networks. In Proceedings of the 2009 ACM Workshop on Cognitive Radio Networks, Beijing, China. New York: ACM; 2009:13-18.

73. Fawaz K, Ghandour A, Olleik M, Artail H: Improving reliability of safety applications in vehicle ad hoc networks through the implementation of a cognitive network. In IEEE 17th International Conference on Telecommunications (ICT). Doha; 4–7 April 2010:798-805.

74. Borota D, Ivkovic G, Vuyyuru R, Altintas O, Seskar I, Spasojevic P: On the delay to reliably detect channel availability in cooperative vehicular environments. In IEEE 73rd Vehicular Technology Conference (VTC Spring). Yokohama; 15–18 May 2011:1-5.

75. Di Felice M, Chowdhury KR, Bononi L: Analyzing the potential of cooperative cognitive radio technology on inter-vehicle communication. In 2010 IFIP Wireless Days (WD). Venice; 20–22 October 2010:1-6.

76. Doost-Mohammady R, Chowdhury KR: Design of spectrum database assisted cognitive radio vehicular networks. In 2012 7th International ICST Conference on Cognitive Radio Oriented Wireless Networks and Communications (CROWNCOM). Stockholm; 18–20 June 2012:1-5.

77. Ghandour AJ, Fawaz K, Artail H, Di Felice M, Bononi L: Improving vehicular safety message delivery through the implementation of a cognitive vehicular network. Ad Hoc Networks 2013, 11(8):2408-2422. 10.1016/j.adhoc.2013.06.005

78. Khabbaz M, Assi C, Ghrayeb A: Modeling and analysis of DSA-based vehicle-to-infrastructure communication systems. IEEE Trans. Intell. Transport. Syst 2013, 14(3):1186-1196.

79. Yang C, Fu Y, Zhang Y, Xie S, Yu R: Energy-efficient hybrid spectrum access scheme in cognitive vehicular ad hoc networks. IEEE Commun. Lett 2013, 17(2):329-332.

80. Chowdhury KR, Felice MD: SEARCH: a routing protocol for mobile cognitive radio ad-hoc networks. Comput. Commun 2009, 32(18):1983-1997. 10.1016/j.comcom.2009.06.011

81. Cacciapuoti AS, Caleffi M, Paura L: Reactive routing for mobile cognitive radio ad hoc networks. Ad Hoc Networks 2012, 10(5):803-815. 10.1016/j.adhoc.2011.04.004

82. Laufer R, Dubois-Ferriere H, Kleinrock L: Multirate anypath routing in wireless mesh networks. In IEEE INFOCOM 2009. Rio de Janeiro; 19–25 April 2009:37-45.

83. Devarapalli V, Wakikawa R, Petrescu A, Thubert P: Network mobility (NEMO) basic support protocol. 2005.

84. Arkko J, Perkins C, Johnson D: Mobility support in IPv6. 2011.

85. Gundavelli S, Leung K, Devarapalli V, Chowdhury K, Patil B: Proxy mobile IPv6. RFC 5213, August (2008)

86. Lee J-H, Ernst T: Lightweight network mobility within PMIPv6 for transportation systems. IEEE Syst. J 2011, 5(3):352-361.

87. Lee J-H, Ernst T, Chilamkurti N: Performance analysis of PMIPv6-based network mobility for intelligent transportation systems. IEEE Trans. Veh. Technol 2012, 61(1):74-85.

88. Lee J-H, Singh KD, Bonnin J-M, Pack S: Mobile data offloading: a host-based distributed mobility management approach. IEEE Internet Comput 2014, 18(1):20-29.

89. De Nardis L, Guirao M-D: Mobility-aware design of cognitive radio networks: challenges and opportunities. In 2010 IEEE Proceedings of the Fifth International Conference on Cognitive Radio Oriented Wireless Networks & Communications (CROWNCOM). Cannes; 9–11 June 2010:1-5.

90. Lee W-Y, Akyildiz IF: Spectrum-aware mobility management in cognitive radio cellular networks. IEEE Trans. Mobile Comput 2012, 11(4):529-542.

91. Wang T, Song L, Han Z: Coalitional graph games for popular content distribution in cognitive radio VANETs. IEEE Trans. Veh. Technol 2013, 62(8):4010-4019. http://dx.doi.org/10.1109/TVT.2013.2245353

92. CORRIDOR: Cognitive radio for railway through dynamic and opportunistic spectrum reuse . Accessed on 11 March 2014. http://corridor.ifsttar.fr

93. Haziza N, Kassab M, Knopp R, Harri J, Kaltenberger F, Agostini P, Berbineau M, Gransart C, Besnier J, Ehrlich J, Aniss H: Multi-technology vehicular cooperative system based on software defined radio (SDR). In NET4CARS 2013, 5th International Workshop on Communication Technologies for Vehicles, Lille, May 14–15, 2013. Heidelberg: Springer; 2013:84-95.

94. Amanna A, Gadhiok M, Price MJ, Reed JH, Siriwongpairat W, Himsoon T: Railway cognitive radio. IEEE Veh. Tech. Mag 2010, 5(3):82-89.

95. CORNET, Cognitive radio network testbed . Accessed on 11 March 2014. http://cornet.wireless.vt.edu

96. Gadhiok M, Amanna A, Price MJ, Reed JH: Metacognition: enhancing the performance of a cognitive radio. In 2011 IEEE First International Multi-Disciplinary Conference on Cognitive Methods in Situation Awareness and Decision Support (CogSIMA). Miami, Beach; 22–24 February 2011:198-203.

97. Amoroso A, Marfia G, Roccetti M, Pau G: Creative testbeds for VANET research: a new methodology. In 2012 IEEE Consumer Communications and Networking Conference (CCNC). Las Vegas; 14–17 January 2012:477-481.

98. Marfia G, Roccetti M, Amoroso A, Gerla M, Pau G, Lim J-H: Cognitive cars: constructing a cognitive playground for VANET research testbeds. In Proceedings of the 4th International Conference on Cognitive Radio and Advanced Spectrum Management Barcelona, 26–29 October 2011. New York: ACM; 2011:29:1-29:5.

99. Open-Access Research Testbed for Next-Generation Wireless Networks (ORBIT). WINLAB, Rutgers University . Accessed on 11 March 2014 http://www.orbit-lab.org

100. Chowdhury KR, Di Felice M, Akyildiz IF: TP-CRAHN: a transport protocol for cognitive radio ad-hoc networks. In IEEE INFOCOM 2009. Rio de Janeiro; 19–25 April 2009:2482-2490.

101. Singh KD, Bonnin J-M, Ros D: Optimising TCP over cognitive radio networks for trains. In IEEE 2012 12th International Conference on ITS Telecommunications (ITST). Taipei; 5–8 November 2012:678-682.

102. Zhang G, Xu Y, Wang X, Tian X, Liu J, Gan X, Yu H, Qian L: Multicast capacity for VANETs with directional antenna and delay constraint. IEEE J. Select. Areas Commun 2012, 30(4):818-833.

103. Singh KD, Ksentini A, Marienval B: Quality of experience measurement tool for SVC video coding. In 2011 IEEE International Conference on Communications (ICC). Kyoto; 5–9 June 2011:1-5.

104. Singh KD, Hadjadj-Aoul Y, Rubino G: Quality of experience estimation for adaptive HTTP/TCP video streaming using H. 264/AVC. In 2012 IEEE Consumer Communications and Networking Conference (CCNC). Las Vegas; 14–17 January 2012:127-131.

105. Deep Singh K, Piamrat K, Park H, Viho C, Bonnin J-M: Optimising QoE for scalable video multicast over WLAN. In 2013 IEEE 24th International Symposium on Personal Indoor and Mobile Radio Communications (PIMRC). London; 8–11 September 2013:2131-2136.

106. Altintas O, Nishibori M, Oshida T, Yoshimura C, Fujii Y, Nishida K, Ihara Y, Saito M, Tsukamoto K, Tsuru M, Oie Y, Vuyyuru R, Al Abbasi A, Ohtake M, Ohta M, Fujii T, Chen Si, Pagadarai S, Alexander Wyglinski M: Demonstration of vehicle to vehicle communications over TV white space (Demo paper). In 2011 IEEE Vehicular Technology Conference (VTC Fall). San Francisco; 5–8 September 2011:1-3.

107. Rzayev T, Shi Y, Vafeiadis A, Pagadarai S, Wyglinski AM: Implementation of a vehicular networking architecture supporting dynamic spectrum access. In 2011 IEEE Vehicular Networking Conference (VNC). Amsterdam; 14–16 November 2011:33-38.

108. SUMO: Simulation of urban mobility. Accessed on 11 March 2014. http://sumo-sim.org

7

CHAPTER

RFID APPLICATIONS: PROSPECTS AND CHALLENGES IN BANGLADESH

Abdul Kadar Muhammad Masum, Faruk Bhuiyan, Kazi Golam Azam

Department of Business Administration, International Islamic University, Chittagong, Bangladesh

ABSTRACT

Radio frequency identification (RFID) is an emerging technology, radio wave communication between a microchip and an electronic reader, consisting of data gathering, distribution, and management systems that has the ability to identify or scan information for remoting recognition of objects with increased speed and accuracy. An attempt has been made to know how using of RFID technology helps to improve services and business process efficiency in public and private sectors of Bangladesh. With this aim, we have conducted extensive literature survey. At the end of this attempt, we have come to the conclusion that the potential use of this technology includes RFID chipped passports, human implants, item-level tagging, inventory tracking and access control systems. RFID technology is at its early stage of adoption in Bangladesh with a few business applications and pilot studies. However, when tags begin to be associated with individuals, privacy is threatened. RFID is a new type of threat to personal information and must be treated as such; indeed, it must be recognized that existing privacy legislation is not adequate. This paper also explores some current and emerging applications of RFID in Bangladesh and examines the challenges including privacy and ethical issues arising from its use, as well as addressing potential means of handling those issues.

Keywords: RFID (Radio Frequency Identification); Potential Applications; Challenges; Bangladesh

1. INTRODUCTION

RFID is not a new technology despite rapidly growing interest in RFID technology in recent years. The concept of the technology dates to the mid to late 1940s, following on from technological developments in the 1930s and the development of radar during World War II [1]. In the 1950s, several technologies related to RFID technology were developed. Technological innovations and their consequences are becoming indispensable parts of our daily lives. RFID, as one of these innovations, is a system that provides easy, secure and quick data entry, storage and transmission. It is used in many places such as shops, stores, hospitals, pharmaceuticals companies, logistic services etc. where real time data should be used. It is important in improving efficiency and visibility, cutting costs, delivering better asset utilization, producing higher quality goods, reducing shrinkage and counterfeiting, and increasing sales by reducing out-of-stocks [2]. Most of developed and developing countries are using RFID technology in their private and public sectors.

Although implementing RFID technology is a complicated process, the right planning and development of an RFID strategy can offer important advantages to business systems and public sectors. While RFID technology has received a fair amount of attention in media recently [3], many are still unfamiliar with RFID and the benefits it can offer. In the face of the need for clear, extensive information about RFID and its benefits, this paper presents the opportunities offered by the technology for any organization or any agency of Bangladesh government involved in the production, management, and sale of goods or services.

In this paper, we motivate our work in Section 2, and the methodology of the research is mentioned in Section 3. Related previous study of RFID Technology is described in Section 4. Furthermore, we have briefly described the working mechanism of RFID technology and its characteristics in Sections 5 and 6. Besides, Section 7 includes RFID applications in present world. Implication of RFID Technology and its possible applications for Bangladesh are explored in Sections 8 and 9.

Moreover, Considerable issues and some suggestions to implementation of RFID technology in Bangladesh are given Sections 10 and 11.

2. OBJECTIVES OF THE STUDY

In this study, an attempt has been made to know how using of RFID technology helps to improve services and business process efficiency in public and private sectors of Bangladesh. However, the specific objectives of the study are set forth as below:

1) To delineate the concise essentials of RFID technology;
2) To explore its current and emerging applications in present world with especial reference of Bangladesh;
3) To evaluate the challenges to implement RFID technology in Bangladesh;
4) To provide some recommendation for prevail over those challenges.

3. METHODOLOGY OF THE STUDY

In light of the objectives of the study, the paper has been designed to investigate how using of RFID technology enhance the efficiency of services and business processes in public and private sector of Bangladesh. To this end an extensive literature survey has been conducted. The study is based largely on secondary data. Data and information from secondary sources were collected by reviewing different published articles, online journals, working papers, existing case studies and websites.

4. LITERATURE REVIEW

Radio frequency identification (RFID) technology is currently one of the technological most promising and discussed auto-identification and data capture (AIDC) technologies [4]. It is not actually a new technology, but it is being applied in new ways, spurred by technological advances and decreased costs. Once used during World War II to identify friendly aircraft, RFID is now being used in a variety of public and private sectors setting from hospital to the highway [5]. A number of studies show that it can be used in many different ways to create value. That is why, several world largest companies use RFID technology to boost up the business performance with great care of consumers and environmental concerns. Chao-Che Lin and Teh-Hsing Ku demonstrates how participating employees with a systemic view can adopt a better methodology to solve problems in RFID and how performance can be enhanced through Systems Thinking [6]. A workshop report was presented from the staff of US federal commission in March, 2005. The workshop brought together RFID proponents, privacy experts, and other interested parties to discuss RFID's various current and potential applications and their implications for consumer privacy. It also highlighted proposals to address these implications and generated discussion about the merits of these different approaches [7]. Another study conducted by Mehmet Barut, Bobert Brown, Nicole Freund, Jonathan may, and Elizabeth Reinhart proposed to add comprehensive ethical and investment deliberation as a means to explore the true value of RFID and its applications [8]. The first exhaustive empirical analysis of performance measurement behavior in the RFID field conducted by to measure the economic performance of their applications and how is the success of RFID applications affected by performance measurement. Nemeth et al. [9] present a state-of-the-art on RFID systems and the challenges and possibilities of the integration to supply chains. Chao et al. [10] review the literature on trends and forecast of RFID technologies from 1991 to 2005 by a historical review method and bibliometric analyze. They focus on the RFID innovation, deployment by enterprises and market diffusion in supply chain management. Recently, Delaunay et al. [11] present a survey on the causes of inventory inaccuracy in supply chain management. Dolgui and Proth [12] also present a literature review on RFID technology in supply chain. They focus on the advantages of RFID technologies in inventory management. They also analyze some problems and present perspectives dealing with privacy and authentication properties of RFID technologies. In Bangladesh, Apollo Hospitals, Dhaka and Bangladesh Army are the early adopters of RFID technology.

5. RFID (RADIO FREQUENCY IDENTIFICATION DEVICE)

RFID has come to signify system solutions for tracking and tracing objects both globally and locally using RFID tags. RFID is one of several technologies collectively known as auto-ID procedures—procedures for identifying objects automatically [13]. RFID by itself is not a location-tracking technology. At sites where readers are installed, RFID may be used to track tagged objects, but this static readability differs from technology such as global positioning systems, or GPS, which uses a network of satellites to pinpoint the location of a receiver [14]. The effectiveness of an RFID application in addressing desired functionality depends on several important factors:

Contactless: An RFID tag can be read without any physical contact between the tag and the reader.

Writable Data: The data of a read-write (RW) RFID tag can be rewritten several times.

Absence of Line-of-Sight: A line of sight is generally not required for an RFID reader to read a tag.

Variety of Reading Ranges: An RFID tag can have a reading range from as small as a few inches to as large as more than 100 feet.

Wide Data-Capacity Range: A tag can store from a few bytes of data to virtually any amount of data.

Support for Multiple Tag Readings: It is possible to use an RFID reader to automatically read several RFID tags in its reading zone within a short period of time.

Durable: RFID tags and readers can easily operate under difficult conditions.

Perform Smart Tasks: In addition to the tasks of carrying and transmitting of data, an RFID system can be designed to perform some other tasks (e.g., acculturation to environmental conditions, operating at high or low temperature and pressure).

Extreme Reading Accuracy: Tanks to extreme reading accuracy advantage, RFID is an accurate and secure technology for data gathering and management.

The advantages listed above are generic for any type of RFID systems. Some additional factors such as privacy and security concerns, data mining, and the integration of the RFID with other technologies such as biometric systems, Global Positioning Systems (GPS) and wireless communication technologies are needed to be considered for applications [15].

6. HOW DOES RFID WORK?

RFID devices have three primary elements: a chip, an antenna, and a reader. A fourth important part of any RFID system is the database where information about tagged objects is stored. The chip contains information about the item to which it is attached. Chips used by retailers and manufacturers to identify consumer goods may contain an Electronic Product Code ("EPC") [16]. EPC chips are encrypted with a unique product code that identifies the individual product to which it is attached, and can be read using radio frequency. The antenna attached to the chip is responsible for transmitting information from the chip to the reader, using radio waves. The chip and antenna combination is referred to as a transponder or, more commonly, as a tag. The reader, or scanning device, also has its own antenna, which it uses to communicate with the tag [17]. RFID systems can employ authentication and encryption to prevent unauthorized reading of data [18]. "Reading" tags refers to the communication between the tag and reader via radio waves operating at a certain frequency. A stationary reader that could be used to track tagged cases of goods entering a warehouse a mobile reader used to monitor inventory on a retail store floor and a prototype of a glove embedded with a scanner used to track daily domestic living activities, a reader communicates with a tag in order to capture the data stored in the tag. The reader usually sends a low-power radio signal to activate the tag and the tag then sends data back to the reader. Most readers are—as their names suggest—only capable of reading data, although some also have the capability to write to certain tags. Normally, readers forward the data to other systems (such as PCs) for subsequent processing. In comparison to tags, readers consume more power, are larger and more expensive. RFID systems use radiosignals to communicate, but only certain frequency bands are available for license-free use. There is international variation in the frequencies and power levels available for RFID systems. Due to different national regulations, an RFID system produced in one country may not work in another country. The database, or other back-end logistics system, stores information about RFID-tagged objects. Access to both a reader and its corresponding database are necessary before information stored on an RFID tag can be obtained and understood.

There are two types of tags, namely active and passive tags. Active tags are larger, heavier and more expensive than passive ones as they contain their own power source (battery). Nevertheless, they emit stronger signal, operating normally at higher frequencies (455 MHz - 950 MHz or 2.45 - 5.8 GHz) capable of being captured by the reader over a longer distance measuring 20 - 100 meters. The tags are more suitable for tracking valuable items. On the hand, passive tags are much smaller, lighter and cheaper than active tags. Typically it is operating at lower frequencies (30 kHz - 134 KHz or 13.6 MHz) with a capture range of between several centimeters and 10 meters. They are commonly used for asset tracking and security applications.

7. PRESENT SCENARIO OF RFID APPLICATIONS IN WORLD

According to the European e-Business Watch large-scale survey [19] of RFID adoption strategies and impacts in four broad economic sectors, 14% of the European companies interviewed were piloting, using or implementing RFID technology in 2007. Companies that were using RFID or planning to use RFID expected major effects on: a) inventory management (49% of companies using or planning to use RFID), b) control and efficiency of inbound logistics (46%), and c) merchandise management and reduced out-of-stocks (44%). The major costs for those using or planning to use the technology were seen to be the costs of project implementation and system integration (39% of all companies using or planning to use RFID). Interestingly, for those companies not using or planning to use RFID technology, 64% stated that a relevant reason for not using it was the insufficient evidence of a strong return on investment (ROI). Wal-Mart and the Department of Defense (DOD) of US told their top suppliers to begin tagging cases and pallets with RFID tags by January of 2005. Indeed, Gartner anticipates that by 2012 most "enterprises will be forced to redesign their value chain processes as a result of RFID changing the storage, collection, and use of data concerning goods in the supply chain". If predictions are true, then the RFID market will approach $4 billion by 2012 [15]. Selected RFID applications in the public sector in some countries are given below:

Health Sector: Austria tests by the municipal administration of Vienna on the applicability of RFID in the health care system. Mexico has Health insurance card: RFID technology is integrated in the "popular insurance" card where the username, information on doctors as well as prescribed drugs are stored. Korea uses RFID technology in hospitals.

e-Passport: In Denmark, e-passport is available since mid-2006; biometric passport relying on RFID embedded fingerprint technology is introduced mid-2009. Germany has introduced e-passport since the end of 2005 and electronic ID card since the end of 2009. Biometric passport is relying on RFID technology in Netherlands, USA and UK. Portugal has e-passport and e-passport control systems at Portuguese airports.

Public Services: Austria tests in the Viennese parking facility management. Germany uses Waste management in different communities. Korea has implemented Pilot projects in the fields of procurement, baggage handling, container management, ammunition management, tracking hazardous waste, museums etc. RFID tags replace paper season parking tickets at car parks in public housing estates in Singapore.

Education Sector: RFID technology is used in Denmark, Germany, Singapore, Netherlands and United States of America implementing for lending systems in libraries.

Logistics/Transport Sector: Japan has set-up of the "Free Mobility Assistance System" based on ubiquitous network technology including RFID tags, to provide information for seamless movement (e.g. transfer routes and transport modes). Netherlands has introduced payment cards for public transport. Singapore establishes Nationwide Electronic Road Pricing (ERP) system to control and manage traffic volume; payment of road usage charges. The ERP is applied to all of Singapore's 840,000 vehicles.

8. IMPLICATION OF RFID TECHNOLOGY FOR BANGLADESH

A little area of both the private and public sectors are driving the RFID market in Bangladesh. Deltech Ltd., a global RFID solution provider has already helped several organizations implement RFID [20]. According to Mr. Delwar Hossain, founder of the Deltech, hands-free access control and employee-tracking systems, as well as asset management are the most popular applications businesses are asking for. Appollo Hospital, Dhaka has completed the first phase of an RFID project to track employee attendance as well as assets [20]. The Bangladesh Army also plans to employ this technology to track soldiers and visitors entering its Dhaka Cantonment as well as track retail purchases. In many export sectors of Bangladesh ranging from garments to household and office furniture, large international buyers can get rid of the age old barcode-based product identification systems for RFID tags to reduce product tracking and inventory management costs of the items which will ultimately help the country to be competitive in the world market. US retailers like Target, Wal-Mart, etc. are major buyers of Bangladesh's garment products. Since Wal-Mart has already announced its plans to require majority of their suppliers to implement RFID based tracking systems, the Bangladeshi exporters have no choice rather to implement RFID technology to survive in the world market. Besides Wal-Mart and Target, German retailer Metro and furniture maker IKEA (recently commenced operations in Bangladesh on a limited basis), to name a few among the large number of companies, have already started experimenting with RFID integrated supply chain management systems. This sea change in global integrated supply chain management necessitates a rapid mobilizetion of our government, IT and export industry leaders and experts to devise a uniform RFID strategy for Bangladesh.

9. POTENTIAL RFID APPLICATIONS FOR BANGLADESH

Radio frequency identification can be used in many different ways to create value and the number is growing at the fast pace. To structure this range of application, some fields of application, which may be suitable for Bangladesh, are given below:

Assets Tracking: RFID can play a vital role to track valuable assets. Companies can put RFID tags on assets that are lost or stolen often, that are underutilized or that are just hard to locate at the time they are needed.

Assets Monitoring and Maintenance: High-value assets can be tagged to store information for maintenance purposes. The tagged machines where the maintenance history and information on replaced parts will be stored on the tag. When data will be stored directly on the tag and not on the companies' network, tags with high data capacity will be needed.

Manufacturing: There is an emphasis on using RFID for applications that can track items from the manufacturing point all the way through to the store shelf. This process entails tagging pallets, cartons, reusable containers and individual items to track the movement of goods throughout the supply chain straight through to the sale of the item to the customer. RFID may be used to help automate the processing the incoming shipments of pistachios from grower partners.

Theft Control: Item level RFID tags may be used to prevent theft along the supply chain or at the point of sale. A simple form is electronic article surveillance (EAS) which can be RFID-based. In this case, low-end RFID systems (e.g. 1-bit tags) will be used which communicate when consumers leave the shop if they have not been deactivated. Applications for theft control in mail order for high-value products such as mobile phones use more sophisticated tags.

Authentication: For authentication purposes, RFID may be used to provide secure identification mechanisms for persons and objects. e-passports combine the traditional paper document with an RFID tag where the critical information is stored. The RFID tag often contains biometric data such as data for facial recognition and fingerprints.

Security and Access Control: RFID may be used as an electronic key to control who has access to office buildings or areas within office buildings. It is convenient (an employee can hold up a badge to unlock a door, rather than looking for a key or swiping a magnetic stripe card) because there is no contact between the card and reader, there is less wear and tear, and therefore less maintenance.

Public Services: Public services include services such as the management of parking facilities and waste management. In waste management RFID may be used for two main purposes: for tracking (hazardous) waste to protect the environment and to allocate costs according to the amount of waste.

Health Service: RFID may be used to track assets such as beds or containers, to identify patients for medication control and to track babies and dementia patients to increase their security. Information such as user name and prescribed drugs are stored on the embedded RFID chip. Hospitals can also plan to deploy RFID to identify patients, call up records, reduce medical errors and improve overall productivity.

Document Administration/Postal Services: RFID technology may be used for the administration of documents. In this field, RFID tags will be attached to documents to improve the location of documents and thus to increase process efficiency and quality. RFID may also be used for postal services in distribution centers to facilitate the sorting of mail items.

Public Transport: RFID may be used by the public sector in the fields of logistics and public transports. One of the most popular uses of RFID may be pay for road tolls without stopping vehicles. RFID is also catching on as a convenient way to pay for bus, subway and train rides. Newer applications are access cards for public transport, RFID-based bus schedules as well as particular location-based services.

10. CHALLENGES TO IMPLEMENTATION RFID TECHNOLOGY IN BOTH PUBLIC AND PRIVATE SECTORS OF BANGLADESH

Utilization of RFID for widespread mass commercial applications has been limited primarily due to challenges arising from several key factors. To assess the overall challenge presented to the organization we propose the use of a simple, “Challenge Assessment Matrix” (CAM) based on Likert-like approach.

Management Commitment: The most significant challenge to implementation is the commitment of management to adopting new technology and having appropriate expectations of RFID capabilities.

Selection of Tag and Tag Reader: Several key factors need to be considered, namely the tag type and size, read distance and other tag reader requirements. Active tags are thicker than the passive ones. Another important feature that should be provided is anti-collision since it allows multiple tags to be read in a single pass.

Cost: Even though the RFID tags are getting cheaper, tag costs are obviously the major variable cost component for RFID, but what about other related cost components? Software, systems integration, process redesign and organizational impacts will be significant and must be part of the business case as well. With the anticipated scale and scope of RFID deployment, tag costs are expected to continue their decline. For RFID implementation, organizations have to analyze Return-on-Investment (ROI) according to their business processes.

Data Management: Most ERP systems and WMS systems are not designed for RFID data capture. The Electronic Product Code occupies a larger data field than the Universal Product Code now in place. New structures for data management should be considered prior to any RFID investment.

International Standards: A key challenge is the continually evolving standards in technology, application, data, conformance, firmware changes, and tracking methods. In addition, different companies often use different standards making cooperation between suppliers and manufacturers difficult.

Availability of Resources: The shortage of existing skilled resources and lack of comprehensive, accessible information has cost implications for training and presents potential implementation problems.

Security: For certain implementations, illicit tracking of RFID tags presents problems. This is particularly relevant for military installations but security challenges are relevant also for corporations and individuals. For example, scanning and cloning of RFID tags can potentially provide undesired access to important facilities or use for payment in commercial transactions.

Business Process Changes: Process automation through RFID will require new work methods and performance measurements for the supply chain. The design of any RFID strategy must consider new processes and procedures to automate tasks and decision making where possible, as well as consider organizational changes.

Ethical Problems: The privacy concern of Bangladeshi consumers is information. The concern of information privacy is the tracking the purchasing habits and other information regarding a particular consumer. This information may be used for targeted advertising or possibly used in evaluating a consumer for various programs such as credit worthiness.

Government Regulations: Implementations are complicated by varying specifications and regulatory requirements, for example operational frequencies and power specifications vary from country to country.

11. RECOMMENDATIONS FOR PREVAIL OVER THOSE CHALLENGES

Bangladesh is always among the last few countries to start using any technology that is already in use for several years in the other developed countries. The primary reasons are the lack of know-how and risk of investment in a new technology. Here, we have provided some suggestions to implement RFID in public and private sectors of Bangladesh.

a) Invest in pilot tests to determine the capabilities and limitations of the software and hardware, as well as the true costs and benefits of RFID. Update the business case based upon the findings.

b) Do not underestimate the need to change business processes to achieve the automation and asset visibility that RFID can provide.

c) At this moment the RFID initiative is discrete in Bangladesh and do not have any long-term strategy leading into implementation-to-diffusion process. Without having a clear strategy to achieve large scale diffusion discrete RFID adoption may not inspire the process in reaching the full potential of RFID.

d) Begin an RFID education effort with a dedicated team that is chartered to recommend and follow a course of action. Research the standards development underway to ensure that the technologies under consideration will meet both domestic and global standards.

e) Businesses must be candid with consumers about the how RFID technology is used in the products which consumers purchase.

f) Consumers should be informed of the existence and location of RFID tag in any product sold. Optimally under these guidelines consumers would have the option of having the devices deactivated at the time of purchase.

g) Ensure that the RFID technology does not interfere with the privacy concerns of consumers is for stores to automatically deactivate RFID tags at point of purchase. RFID tags can be killed in a number of ways; tags designed with a kill switch built in so that upon receiving an instruction they deactivate themselves permanently.

h) Consumer education is a vital part of protecting consumer privacy. Industry members, privacy advocates, and government should develop education tools that inform consumers about RFID technology, how they can expect to encounter it, and what choices they have with respect to its usage in particular situations.

i) Many of the potential privacy issues associated with RFID are inextricably linked to database security. As in other contexts in which personal information is collected from consumers, a company that uses RFID to collect such information must implement reasonable and appropriate measures to protect that data.

j) Government should regulate the use of RFID tags in consumer goods. Optimally a law would require that RFID tags be permanently disabled at the time of purchase. The state would have in the usefulness of RFID tags in public activates such as law enforcement, it is unlikely that the government will be quick to set controls on how these devices are used.

k) Govern should design of an Auto-ID strategy for Bangladesh with a focus upon how business processes and the customer experience can be improved.

12. CONCLUSION

Despite the hype, the promise of RFID technology is real, and can provide increased asset visibility, enhanced information content and velocity. Equally as real are the pitfalls of RFID including cost, competing standards, technology limitations and privacy concerns. The transition to RFID may not occur overnight, but this technology has been given a jump start by leading industries in Europe. Bangladesh can adopt any suitable model for its overall automation in public and private sectors. Anticipate that return on investment for RFID will be low or non-existent initially, but that future gains are possible by redesigning business processes. We recommend RFID education for Bangladeshi consumers and development of the business case and pilot to size the potential benefits and adoption efforts. Finally, we recommend the design of an Auto-ID strategy for Bangladesh with a focus upon how business processes and the customer experience can be improved.

REFERENCES

1. Danish Board of Technology, "RFID from Production to Consumption—Risks and Opportunities from RFID Technology in the Value Chain," 2006. www.tekno.dk/pdf/projekter/p06_rapport_RFID.pdf
2. C. C. Chao, J. M. Yang and W. Y. Jen, "Determining Technology Trends and Forecasts of RFID by a Historical Review and Bibliometric Analysis from 1991 to 2005," Technovation, Vol. 27, No. 5, 2007, pp. 268-279.
3. OECD, "RFID Implementation in Germany: Challenges and Benefits," 2007. www.oecd.org/dataoecd/19/23/39693586.pdf
4. M. Martin, "The History of RFID Technology," RFID Journal, 2007. http://www.rfidjournal.com/article/articleview/1338/1/129/
5. C. Delaunay, E. Sahin and Y. Dallery, "A Literature Review on Investigations Dealing with Inventory Management with Data Inaccuracies in RFID Eurasia, 1st Annual," 2007, pp. 1-7.
6. W. Minli, W. Gang and H. Dajun, "Device Management in RFID Public Service Platform," 3rd International Conference on Convergence and Hybrid Information Technology, Busan, Vol. 1, 2008, pp. 1133-1136.
7. Y. Choi, D. Kang, H. Chae and K. Kim, "An Enterprise Architecture Framework for Collaboration of Virtual Enterprise Chains," The International Journal of Advanced Manufacturing Technology, Vol. 35, No. 11-12, 2008, pp. 1065-1078.
8. T. Phillips, T. Karygiannis and R. Kuhn, "Security Standards for the RFID Market," IEEE Security and Privacy, 2005, pp. 85-88.
9. A. Dolgui and J. M. Proth, "RFID Technology in Supply Chain Managment: State of the Art and Perspectives," 17th World Congress the International Federation of Automatic Control, Seoul, 2008, pp. 4465-4475.
10. A. Sarac, N. Absi and S. Dauzere-Peres, "A Literature Review on the Impact of RFID Technologies on Supply Chain Management," International Journal of Production Economics, Vol. 128, No. 1, 2010, pp. 77-95.
11. S. Veronneau and J. Roy, "RFID Benefits, Costs, and Possibilities: The Economical Analysis of RFID Deployment in a Cruise

Corporation Global Service Supply Chain," International Journal of Production Economics, Vol. 122, No. 2, 2009, pp. 692-702.

12. V. D. Hunt, A. Puglia and M. Puglia, "RFID: A Guide to Radio Frequency Identification," John Wiley & Sons Inc., New Jersey, 2007.

13. Y. Xiao, S. Yu, K. Wu, Q. Ni, C. Janecek and J. Nordstad, "Radio Frequency Identification: Technologies, Applications," Wireless Communications and Mobile Computing, Vol. 7, No. 4, 2007, pp. 45-472.

14. B. Bacheldor, "RFID Takes Root in Bangladesh," RFID Journal, 2008.

15. P. Nmeth, L. Toth and T. Hartvanyi, "Adapting RFID in Supply Chains," 3rd International Conference on Mechatronics, 2006, pp. 263-266.

16. M. A. Khan, M. Sharma and R. B. Prabhu, "A Survey of RFID Tags," International Journal of Recent Trends in Engineering, Vol. 1, No. 4, 2007, pp. 68-71.

17. Y. Choi, D. Kang, H. Chae and K. Kim, "An Enterprise Architecture Framework for Collaboration of Virtual Enterprise Chains," The International Journal of Advanced Manufacturing Technology, Vol. 35, No. 11-12, 2008, pp. 1065-1078.

18. M. Ohkubo, K. Suzuki and S. Kinoshita, "RFID Privacy Issues and Technical Challenges," Communications of the ACM, Vol. 48, No. 9, 2005, pp. 66-71.

19. M. L. Chuang and W. H. Shaw, "RFID: Integration Stages in Supply Chain Management," IEEE Engineering Management Review, Vol. 35, No. 2, 2007, pp. 80-87.

20. S. B. Miles, S. E. Sarma and J. R. Williams, "RFID Technology and Applications," Cambridge University Press, Cambridge, 2008.

8
CHAPTER
PASSIVE UHF RFID TAG WITH MULTIPLE SENSING CAPABILITIES

José Fernández-Salmerón [1,†], Almudena Rivadeneyra [1,†], Fernando Martínez-Martí [2], Luis Fermín Capitán-Vallvey [3], Alberto J. Palma [2] and Miguel A. Carvajal [2,*]

[1] *The Institute for Nanoelectronics, Technical University of Munich, DE-80333 Munich, Germany*

[2] *Department of Electronic and Computer Technology, ECSens, University of Granada, E-18071 Granada, Spain*

[3] *Analytical Chemical Department, ECSens, University of Granada, E-18071 Granada, Spain*

ABSTRACT

This work presents the design, fabrication, and characterization of a printed radio frequency identification tag in the ultra-high frequency band with multiple sensing capabilities. This passive tag is directly screen printed on a cardboard box with the aim of monitoring the packaging conditions during the different stages of the supply chain. This tag includes a commercial force sensor and a printed opening detector. Hence, the force applied to the package can be measured as well as the opening of the box can be detected. The architecture presented is a passive single-chip RFID tag. An electronic switch has been implemented to be able to measure both sensor magnitudes in the same access without including a microcontroller or battery. Moreover, the chip used here integrates a temperature sensor and, therefore, this tag provides three different parameters in every reading.

Keywords: screen printing; printed electronics; UHF antenna; RFID tag; sensor; switch

1. INTRODUCTION

One emerging technology to control the supply chain is radio frequency identification (RFID). The vast majority of RFID tags consist of an antenna and an integrated circuit (IC). Most RFID tags are low-cost passive tags, which take advantage of the reader-induced signal to supply power to the RFID chip [1]. In this sense, printed electronics results in a good alternative to produce low-cost systems in large-scale production. Several examples are found of RFID tags developed with different printing techniques such as screen printing, gravure, flexography, or inkjet printing [2,3,4,5]. In addition to cost savings, printed electronics can produce thin, wearable, and lightweight devices, as well as flexible systems by using malleable substrates [6].

In recent years, there has been a growing interest in developing RFID tags with sensing capabilities. Different approaches have been followed to achieve this goal. On one side, some authors have studied the analog response of the sensing RFID tag: read range, shift of the antenna resonance frequency, measured level of backscattered power, etc. Changes in these electrical parameters have been associated to a variation of the monitored magnitude [2,7,8,9,10,11,12,13,14,15,16,17,18], resulting in useful strategies as threshold sensor tags [19] but they lead to uncertainties introduced by the wireless link between RFID reader and tag. It is difficult to know whether a change in the analog response of the tag can be directly related to a variation in the monitored magnitude or it is affected by other factors that interfere in the link performance, such as path loss between antennas, spatial and time drifts, electrical interference, non-direct sight between reader and tag, interferences with other objects, etc. Furthermore, extra-circuitry typically has to be included into the RFID reader to monitor the tag analog response. On the other hand, other approaches are based on microcontroller architectures in combination with RFID chips and different types of sensors: temperature, light, and moisture content [20,21]; chemical sensing [22,23,24,25]; printed moisture sensors [26,27]; pressure [28,29]; general RFID platforms for different sensing applications [30,31], SAW sensors [32] or built-in sensors, commonly temperature sensors [33]. Additionally, some examples of single-chip architecture without microcontroller units have been reported by [34,35,36,37]. The main advantage of this strategy compared to the analog approach is the processing of the sensor data directly in the RFID tag. Thus, the sensor data, in digital form, can be reliably delivered to the reader or stored on the tag for future access. Nevertheless, these strategies require the use of a battery to enable data logging and power the necessary extra-chips, such as a microcontroller unit to drive the different sensors. All these extra components increase the cost of the system.

Here, we present the design, fabrication and characterization of a printed RFID tag with sensing capabilities. This passive tag is printed on a cardboard box with the aim of checking the packaging conditions. The tag includes a commercial force sensor and a printed opening detector. In this sense, the mass applied to the packaging can be measured as well as the detection of the opening of the box. These two sensors provide relevant information during all stages of the supply chain. For example, shock events and applied forces invariably occur in the shipping industry but they are difficult to be detected by different stakeholders. For instance, a study showed that a package would experience, on average, a drop from a height of 1–1.85 m during its transit [38] and another study found that an average commercial truck would induce loads from 5–35 g [39,40]. Sensors could be mounted on shipping containers or packages to determine if any force has been applied in its transit through the supply chain [41].

As remarkable features, the architecture presented here is based on a single-chip architecture using screen printing technology as fabrication process. In order to use a unique chip with two sensors, an electronic switch has been included in the tag that reads them independently. Thus, both sensor magnitudes can be measured in the same access without including a microcontroller or battery. Our tag only requires several external transistors and resistors to obtain sensor data. Due to the passive architecture, it is necessary to power up the tag from the electromagnetic field radiated by the RFID reader. Obviously, the opening detector has memory (once it is broken, it is not possible to recover it), whilst the force pressure sensor provides information about the mass applied at the moment of the sensor reading. Moreover, due to the fact that the chip used here integrates a temperature sensor, these tags provide three parameters in every reading. Regarding the frequency band, the ultra-high frequency band (UHF) has been chosen here for its higher read range and the possibility of simultaneous detection of several tags. The high frequency (HF) band could have interesting results with the implementation of the near-field communication (NFC) protocol that many mobile phones already include, and therefore no specific reader would be required. The drawback of this band is its shorter read range, no more than few centimeters.

2. MATERIAL AND METHOD

2.1. Smart Tag Architecture

Two resistive sensors are integrated in this tag: a force sensor and an opening detector. In order to measure the vertical force applied to the packaging, we selected the FlexiForce A201 force sensor (Tekscan, Inc., South Boston, MA, USA) working in the force range up to 111 N (25 pounds). This sensor presents an error lower than 3% and its hysteresis is lower than 4.5%. Its time response is below 5 μs. As an opening detector, we printed a conductive pattern through the die cut line of the package. In this sense, we are able to measure a low resistance value when the package is closed, whereas when it is opened, this printed pattern is broken and, therefore, an open circuit is detected.

Temperature value comes from a conversion in the on-chip A/D converter of the SL900A. Two internal voltage references, V_{ref1} and V_{ref2} individually selectable in steps of 50 mV between 160 and 610 mV, set the lower and upper limits of this converter. These limits are defined as $2V_{ref1}-V_{ref2}$ and V_{ref1}. The difference between them defines the input voltage range, $V_{ref2}-V_{ref1}$, and the limits of operation. These voltage references can be set in the user application and, therefore, a concrete resolution and range can be selected by the user. The minimum resolution of the built-in temperature sensor is 0.18 °C in a range from −89.3 °C to 84.6 °C, while a resolution of 0.23 °C is obtained with the widest range from −89.3 °C to 147.9 °C. It should be noted that operation of the RFID chip below −40 °C and above 125 °C is not guaranteed by the manufacturer.

The resistive sensors are directly connected to the sensor front end (SFE) of the RFID chip. To read out a resistance value, it does not require any external components but, in our case, we want to measure sequentially two different resistive sensors in the same access. For this purpose, a switch based on discrete MOS transistors has been implemented (see Figure 1). The transistors are BSS138LT1 NMOS (Semiconductor Components Industries LLC, USA) with a resistance value in on-region $R_{DS(on)}$ = 3.5 Ω and SMD pull-up resistors of 4.7 kΩ (Panasonic, Japan). V_{EXC} is the switch control signal that is controlled by the RFID reader. V_{EXC} terminal, which is usually employed as voltage supply for external sensors, can be set to high or low level. In this application, V_{EXC} is used as the switch control signal:

when V_{EXC} is set at the low level, the sensor R_{Sens1} is selected, whereas R_{Sens2} is measured when high level is set in V_{EXC} (Figure 1). Therefore, only one sensor is read each time. Two inverters (Q3 and Q4) are needed in order to ensure the maximum available voltage in the gates of Q1 and Q2 and reduce the ON resistance, $R_{DS(on)}$, of these transistors. In fact, the measured $R_{DS(on)}$ value is only a few Ω that is negligible in comparison with the sensors resistance, as will be shown below.

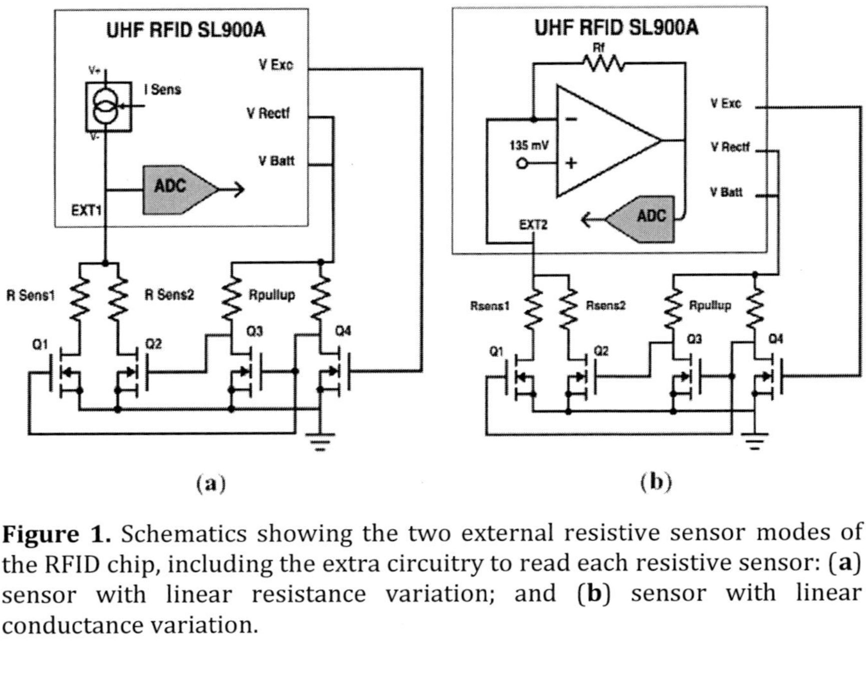

Figure 1. Schematics showing the two external resistive sensor modes of the RFID chip, including the extra circuitry to read each resistive sensor: (**a**) sensor with linear resistance variation; and (**b**) sensor with linear conductance variation.

Figure 2 shows the layout of the printed tag, including the footprints of the different components required. This tag presents a passive architecture based on SL900A RFID chip (AMS AG, Unterpremstaetten, Austria) compatible with EPC Gen 2 RFID standard [42]. The RFID chip was chosen due to its integration of a SFE that comprises different sensor conditioning stages and a 10-bit analog-to-digital converter (ADC). Extra circuitry has been used to interface the resistive sensors. In addition, this RFID chip includes an in-built temperature sensor. The radio frequency interface consists of a typical dipole antenna resonating at 868 MHz (European UHF RFID band) and a RF surface mount device (SMD) inductor used to match the chip input impedance [43]. The dipole antennas are the best candidate for RFID applications because of its nearly omnidirectional character, i.e., labels can be detected in almost any relative position of the RFID reader to the antenna. In case of choosing a different frequency band, it would be only necessary to change the antenna layout and the RFID chip.

In order to read out the smart tag by a RFID reader, several EPC Gen 2 custom commands must be sent to select V_{EXC} level and different parameters to read each resistive sensor. In this particular case, Set_SFE_Parameters command is employed to choose the SFE configuration and Set_Calibration_Data command is sent to configure the ADC reference and V_{EXC} signal. Another EPC Gen 2 command, Get_Sensor_Value, is used then to get the resistance value. EPC Gen 2 command delay must be lower than 20 ms according to protocol specifications. This delay depends on the selected data rate (5–640 kbits/s); forward and backward links operation between RFID reader and tag and the number of tags detected by the reader. An extra time of 6 ms is employed by the ADC converter to acquire and process the resistance values in the case of Get_Sensor_Value command. Due to the fact that this process is realized sequentially to read out each resistive sensor, the total time required, assuming the worst case (maximum delay between reader and tag), is below 66 ms.

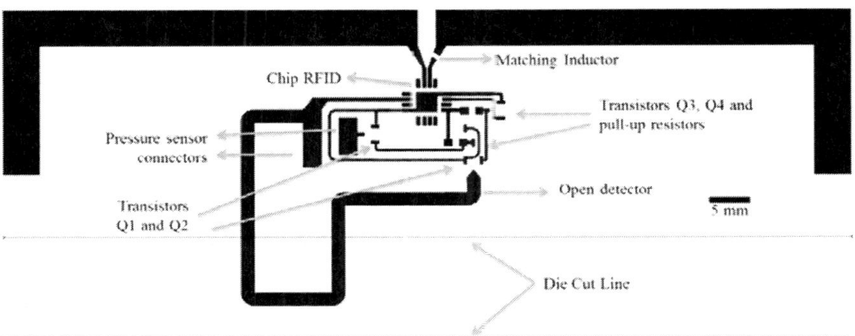

Figure 2. Layout of the RFID tag comprising a dipole antenna, an RFID chip and the sensor architecture. Components have been labeled for better understanding.

Two different SFE configurations are used to read out the resistive sensors: linear resistance and linear conductance. The former configuration (Figure 2a) is indicated for sensors whose resistance varies linearly with the sensed magnitude. The sensor resistance selected by the switch is polarized with a configurable current source (I_{SENS}) whose amplitude can be selected from 0 to 7.75 μA, in steps of 250 nA. The input voltage of the ADC, V_{ADC}, is associated with the resistance value as follows:

$$V_{ADC} = I_{SENS}(R_{SENS,i} + R_{DS(on)}) \tag{1}$$

In order to measure both sensors, it is necessary to properly select the current value; the opening detector was measured with a current equal to I_{SENS} = 7 μA, whilst the force sensor was configured with I_{SENS} = 6 μA. This difference in current is given by the range of resistance variation; in the opening detector case, it is enough to detect a change from a few Ω (close state) to open circuit (open state). In the force sensor case, the resistance value, which varies from some MΩ (no pressure) to a few kΩ, is measured to determine the applied force. As previously explained, the limits of the ADC input voltage are V_{ref1} and $2V_{ref1}–V_{ref2}$. In the case of the opening detector, these ADC references are set to V_{ref1} = 260 mV and V_{ref2} = 0 V, so the ADC input voltage ranges from 260 mV to 520 mV leading to a detection range of resistance from 37 to 74 kΩ. For the force pressure sensor, the ADC references are set to V_{ref1} = 610 mV and V_{ref2} = 0 V, giving an ADC input voltage range of 610–1220 mV that corresponds to a detection range of 100–203 kΩ for a sensing current of 6 μA. Any other detection range could be easily configured by choosing the appropriate sensing current and ADC limits.

The latter readout configuration (Figure 2b) is suitable for sensors whose conductance changes linearly with the sensed magnitude. This is the case of the force sensor used in this work. An operational amplifier with negative feedback is combined with a selectable feedback resistor R_{FB} (185, 400, 875, 1875 or 3875 kΩ), see Figure 2b. The non-inverting input is fixed to 135 mV, so the input voltage in the ADC is related to R_{SENS} value:

$$V_{ADC} = 135\ mV + 135\ mV\ \frac{R_{FB}}{R_{SENS,i}} \tag{2}$$

For optimal operation, the feedback resistor value and the ADC reference voltages have been chosen as close as possible to the resistive sensor values and the non-inverting input voltage, respectively: the feedback resistor is 185 kΩ whilst V_{ref1} is 260 mV and V_{ref2} is set to 0 V. This configuration leads to a detection range for the resistive sensors from 65 to 200 kΩ.

2.2. Fabrication Process

All the patterns were directly printed on a commercial cardboard box (Sociedad Estatal Correos y Telégrafos, S.A., Madrid, Spain) with dimensions 317 × 215 × 125 mm and thickness of 1.5 mm. Screen printed patterns were manufactured with a Serfix III screen printing machine (Seglevint SL, Barcelona, Spain). The screen used to manufacture the tag had a mesh count of 120 Nylon thread per centimeter (T/cm). The prototype was fabricated with only one screen printed layer of conductive silver ink CRSN 2442 (Sun Chemical Corporation, Parsippany, NJ, USA). The minimum spatial resolution for a reliable fabrication process was 150 μm. Sintering took place at a constant temperature of 80 °C for 20 min [43].

Finally, the assembly of the chips and external components to the cardboard box was carried out with a three-step process. First, the interconnections between RFID chip and silver pads were made by using the conductive resin H20E (Epoxy Technology, Inc., Billerica, MA, USA). A double layer of 50 μm-thick dry adhesive, AR Clear 8932 (Adhesives Research, Inc., Glen Rock, NJ, USA), was placed on the bottom of the chip to fix it to the substrate. Finally, the conductive resin was cured by heating in an oven at 120 °C for 20 min. Moreover, the dry film adhesion improved with temperature, so this heat treatment served also to better fix the components to the substrate.

3. RESULTS AND DISCUSSION

3.1. Antenna Characterization

Figure 3 presents the manufactured RFID tag with its different components labeled for better understanding. Final dimensions of the screen printed dipole antenna arms are 5.5 mm width and 72 mm length. The thickness and roughness of the printed layers were measured with the Dektak XT™ Stimulus Surface Profiling System (Bruker Corporation, Conventry, UK), finding a thickness of 17 ± 2 μm with a root mean square (RMS) roughness of 1.7 μm. In order to optimize the occupied area, the dipole arms have been bent. The input impedance of the dipole is (31.1 + j9.4) Ω at 868 MHz. The impedance of this dipole has been designed to achieve the same real part as the one of the RFID chip, whilst the necessary large inductive part is achieved with a SMD inductor series 3650 (TE Connectivity, Ltd., Schaffhausen, Switzerland) placed on one of its arms. Its inductance is 51 nH and its quality factor is 60 at 900 MHz. The antenna parameters have been obtained by EM simulation with Advanced Design System 2013 (Keysight Technologies Inc., Santa Rosa, CA, USA). Substrate influence has been included in the simulations and its electrical parameters have been taken from [31]. In particular, its gain is 0.661 dBi, its directivity 2.401 dBi and its efficiency 66.97% at the working frequency [29].

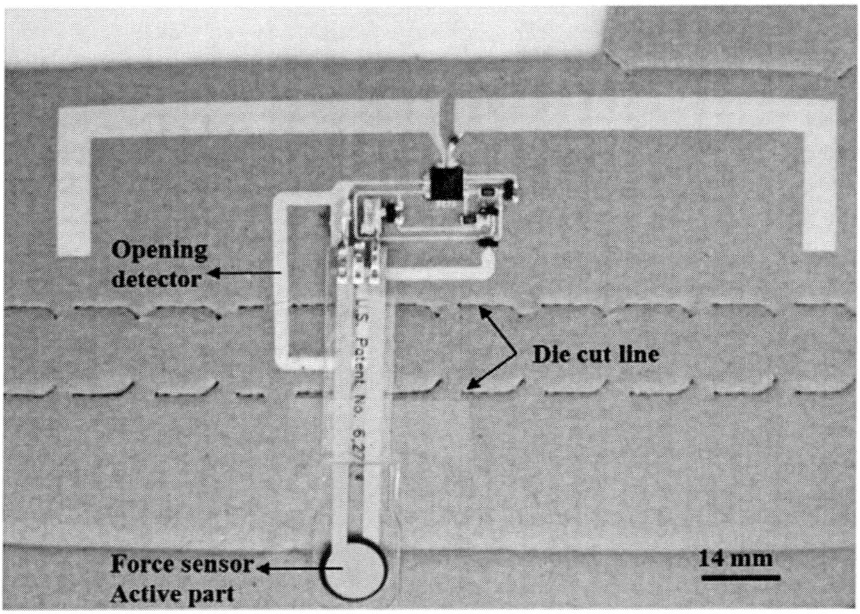

Figure 3. Fabricated RFID tag with all components.

According to the theory of communication in RFID systems [43,44], the maximum read range is calculated using Equation (3).

$$range_{max} = \frac{\lambda}{4\pi}\sqrt{\frac{G_{tag}G_{reader}P_{reader}\tau\, PLF}{S_{tag}}} \tag{3}$$

where S_{tag} is the RFID chip sensitivity; the minimum received power level to activate the tag is −15 dBm for the chosen RFID chip. G_{tag} refers to the tag antenna gain, P_{reader} is the effective radiated power by the reader and G_{reader} the reader antenna gain. The polarization mismatch between the reader antenna (circular polarization) and the designed tag antenna (lineal polarization) is also included. This is represented as the path loss factor (PLF) in Equation (3) and its value is 0.5 (3 dB). The transmission power is 26 dBm and the reader antenna gain is 7 dBi at 868 MHz, according to the manufacturer. The factor τ takes into account the losses related to the mismatch between the chip and the antenna [43].

The measurements of the maximum read range took placed on anechoic chamber; the antenna reader was placed on a tripod whilst the RFID tags printed on the cardboard box were placed at the same height. RFID tag tests were performed using a commercial RFID reader compatible with EPC Gen 2, DK-UHF RFID HP2 (IDS Microchip AG, Wollerau, Switzerland). The read range is the maximum distance in which the reader is able to read out the tag including the sensing functionalities. It should be noted that the sensitivity of the RFID chip has a penalty. The chip sensitivity to answer identification inquiries of EPC protocol is −15 dBm, whereas the sensitivity to read out a value of resistance is −2.21 dBm. The reason to explain these differences is the fact that extra power is required to drive the SFE. This extra power is collected from the radiated EM field of the reader. This aspect has been already studied in our previous works [29,37] and it is calculated using the antenna parameters and on-chip RF rectifier efficiency, ~40% according to the manufacturer.

According to Equation (3) and setting chip sensitivity to −2.21 dBm, the maximum read range should be 2.2 m, assuming ideal conditions (perfect matching between chip, $\tau = 1$, and the dipole antenna gain obtained by EM simulation, $G_{tag} = 0.66$ dBi). However, the measured range for the printed tag at room conditions is smaller than the simulated one, 1.1 m. These differences between simulated and measured read ranges were already reported on our previous work [29], being attributed to the following factors: the non-ideal behavior of the metallic layers in printed electronic that leads to a lower gain and efficiency of the fabricated antennas [29] and extra losses introduced by the mismatch between the impedances of RFID chip and antenna because of the variation of the RFID chip impedance with both working frequency and powers levels In this case, the read range has been reduced approximately two thirds compared with the same antenna fabricated on polyimide substrate [29]. Although the thick cardboard substrate influence has been taken in account, this reduction could be associated with extra losses and detuning of the RFID antenna operating on paper substrate [31]. Moreover, there is a non-direct sight between RFID reader and tag antenna because the whole tag is printed on the inner side of the cardboard box. To justify this reduction, a commercial UHF RFID Raflatac tag from UPM RFID (SMARTRAC N.V., Amsterdam, The Netherlands) has been detected on the same conditions: a direct line of sight to the RFID reader on free air condition and attached to the inner side of the cardboard box. In these scenarios, the measured read ranges were 3.3 m and 2.28 m, respectively. In this case, the read ranges are larger due to the lower threshold voltage to activate these RFID tags, but the read range is penalized in the same way as our dipole antenna explaining the results found.

3.2. Sensing Tag Performance

Figure 4 shows the cardboard box with the printed RFID tag. Regarding the force sensor, its active part has been placed just in the junction part of the cover and lateral of the box (Figure 4b). When some weight is applied to the top of the box, the pressure force is distributed among the junction between the cover and lateral sides of the box. Anyway, the force sensor can be placed where the manufacturer considers more critical. With respect to the opening detector, it has been defined through the die cut line in such a way that once the box is closed, to open it again, it is necessary to strip the die cut line off. Therefore, the opening detector is obviously non-reversible.

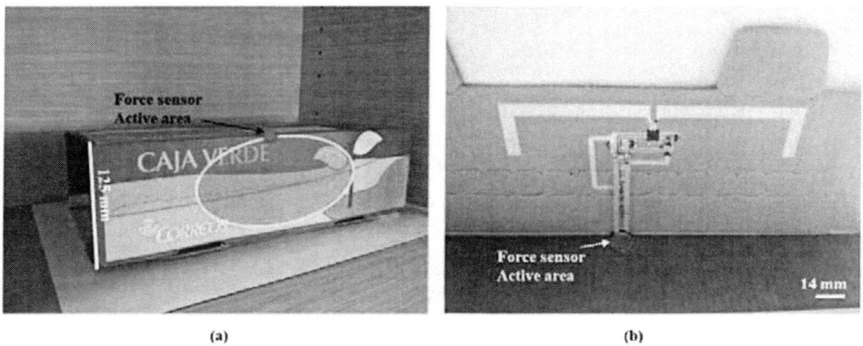

Figure 4. (**a**) Close cardboard with RFID tag; (**b**) view of the RFID tag before closing the box.

Temperature measurements are directly obtained from the on-chip sensor and present different errors and accuracy depending on the selected measurement range as mentioned above.

After setting the different configurations, we have characterized the opening detector and force pressure sensor using the two presented configurations: linear resistive and linear conductive modes (see Figure 2a,b, respectively), as well as the in-built temperature sensor.

Regarding the opening detector in linear resistive mode, the maximum ADC counts are read when the packaging is opened. On the contrary, the zero level is returned while the box is closed. No error was observed in any of the performed test. A similar behavior has been found in the linear conductance mode, being the maximum ADC counts achieved in the case of a closed package and zero when the package has been opened. This behavior is expected due to the fact that the two resistance values (a few Ω and open circuit) are out of ADC limits, in both cases.

Regarding the force sensor, a mass has been applied to the top of the force sensor. Mass has been measured with a weighing scale model Soehnle 67080 (Leifheit AG, Nassau, Germany) with a resolution of 1 g. We have stored and averaged fifteen ADC data to calculate the deviation between measurements. This deviation has been lower than 15 Least Significant Bit (LSB) of the ADC in all the studied cases. Figure 5 presents the average data obtained in the ADC for the force sensor on both operation modes. The output range for the force sensor is equal to the maximum resolution of the internal ADC, that is to say, 1024 counts can be associated with a variation of about 4.17 kg in the first SFE configuration (Figure 5a), whereas this output range in the second configuration (Figure 5b) corresponds to a change of about 5.27 kg. This force sensor presents a resistance of some MΩ when there is no mass on it, decreasing its value while increasing the mass. Detection limits are different due to the different detection ranges of each configuration. In the first case, we have been able to discriminate values from 1 kg to 4 kg, whilst in the second case, the lower limit is 1.2 kg and the upper limit is 5 kg. With respect to the hysteresis of the measurements, the maximum error found in both configurations is about 3%, which is within the error of the force sensor.

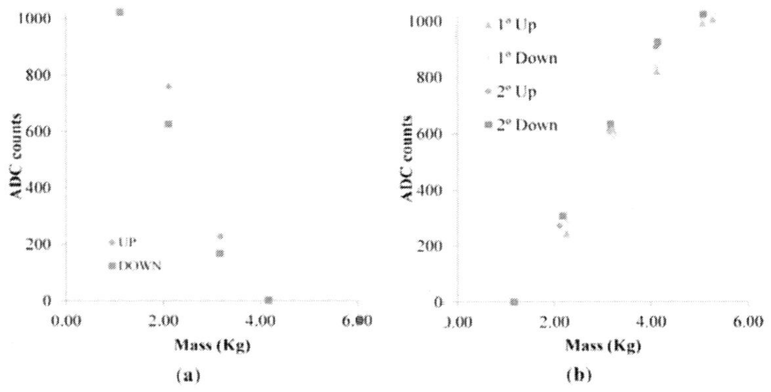

Figure 5. ADC counts for the force sensor measurements: (**a**) configuration as linear resistance; and (**b**) configuration as linear conductance.

The conductive epoxy used in this work has a resistivity below 0.4 mΩ·cm and was deposited using a 140 T/cm screen that gets ~10 μm thickness, see [45], leading to a sheet resistance of roughly 4 Ω/□. Screen silver patterns gets a resistivity in the range of 40–50 mΩ·cm or, equivalently, a sheet resistance of 0.02–0.03 Ω/□. Although this difference is important, the epoxy has been used only in the chip pins (size of 0.5 mm width and 1.4 mm length), pressure sensor interconnection (2.6 mm width and 5.6 mm length), and switch transistors pads (1.2 mm width and 0.4 mm length). Using the calculated epoxy sheet resistance and taking into account that there are several connections for each resistive sensor; the increase in resistance is about 26 Ω for the opening detector sensor and about 42 Ω for the force sensor. In the case of the force sensor, this increment in resistance is neglected compared to its intrinsic resistance. In the case of the opening detector sensor, this increment in resistance is not critical because the sensor changes from several kΩs to open-state condition. The on-chip sensor front end and ADC are powered internally from the internal rectifier so epoxy resin only affects to antenna–RFID chip connections. This aspect has been fully studied in [29] and RFID chip input impedance was measured using the epoxy as the bonding material, so its influence has been taken into account and fully studied in [29].

Finally, the tag was introduced in the climatic chamber VCL 4006 (Vötsch Industrietechnik GmbH, Balingen, Germany) and the temperature was programmed to decrease from 30 °C to 20 °C in 10 min at RH room conditions. The temperature measurement difference between the chamber sensor and tag sensor always was below 0.5 °C, which is covered by their respective uncertainties.

4. CONCLUSIONS

This work describes a dipole-based UHF RFID tag with multiple sensor capabilities integrated on a cardboard package. This tag has been manufactured by screen printing without any pre-treatment of the cardboard. A simple conditioning circuit has been designed and tested to allow the multiplexing of one SFE input. This solution improves the sensing capability of this kind of RFID chip. In this work, two resistance sensors are included in the design: a silver trace crossing the die cut line of the package to detect if the package has been opened and one force sensor to measure how much weight is applied to the top of the package. In particular, the selected force sensor is able to discriminate weights in the range of 1 to 5 kg, depending on the sensing configuration mode and selected parameters.

Furthermore, the chosen RFID chip includes an in-built temperature sensor. Therefore, three different magnitudes can be measured in every reading with neither microcontroller nor battery.

In order to connect both sensors to the single-sensor interface of the RFID chip without including a microcontroller, a switch has been integrated on this design. Four transistors compose this switch: two transistors to control which resistive sensor is accessible each time and two inverters to ensure the maximum available voltage in the gates of the control transistors.

The operation of this passive tag has been successfully demonstrated. Two different modes of acquiring resistance values have been tested: linear conductance and linear resistance. The opening detector works well in both configurations. Regarding the force sensor, the linear conductance configuration results are more appropriate because a larger detection range, better linearity, and lower hysteresis are achieved compared with the linear resistance configuration. This passive tag would be very useful to detect the package conditions during the different stages of the supply chain where an RFID reader is available.

ACKNOWLEDGMENTS

This work was partially funded by the Ministerio de Educación y Ciencia under Projects CTQ2009-14428-C02-01 and CTQ2009-14428-C02-02 and the Junta de Andalucía (Proyecto de Excelencia P10-TIC-5997), Spain. This project was partially supported by European Regional Development Funds (ERDF).

AUTHOR CONTRIBUTIONS

José Fernández-Salmerón and Almudena Rivadeneyra developed the design of the tag and its characterization. Fernando Martínez-Martí provided the background and contributed to the experimental characterization of the pressure sensor. Luis Fermín Capitán-Vallvey conceived and supervised the screen printing process, Alberto J. Palma supervised the electronic design and the sensor selection and Miguel A. Carvajal designed the switch topology related to the multisensing capability of the tag. All authors contributed to the discussion of the results and the manuscript writing.

CONFLICTS OF INTEREST

The authors declare no conflict of interest.

REFERENCES

1. Finkenzeller, K.; Müller, D. RFID Handbook: Fundamentals and Applications in Contactless Smart Cards, Radio Frequency Identification and Near-Field Communication; Wiley: New York, NY, USA, 2010.
2. Virtanen, J.; Ukkonen, L.; Bjorninen, T.; Elsherbeni, A.Z.; Sydanheimo, L. Inkjet-Printed Humidity Sensor for Passive UHF RFID Systems. IEEE Trans. Instrum. Meas. **2011**, 60, 2768–2777.
3. Sangoi, R.; Smith, C.G.; Seymour, M.D.; Venkataraman, J.N.; Clark, D.M.; Kleper, M.L.; Kahn, B.E. Printing radio frequency identification (RFID) tag antennas using inks containing silver dispersions. J. Dispers. Sci. Technol. **2005**, 25, 513–521.
4. Allen, M.; Lee, C.; Ahn, B.; Kololuoma, T.; Shin, K.; Ko, S. R2R gravure and inkjet printed RF resonant tag. Microelectron. Eng. **2011**, 88, 3293–3299.
5. Martínez-Olmos, A.; Fernández-Salmerón, J.; Lopez-Ruiz, N.; Rivadeneyra Torres, A.; Capitan-Vallvey, L.F.; Palma, A. Screen printed flexible radiofrequency identification tag for oxygen monitoring. Anal. Chem. **2013**, 85, 11098–11105.
6. Suganuma, K. Introduction to Printed Electronics; Springer Science & Business Media: Berlin, Germany, 2014; Volume 74.
7. Potyrailo, R.A.; Nagraj, N.; Tang, Z.; Mondello, F.J.; Surman, C.; Morris, W. Battery-free radio frequency identification (RFID) sensors for food quality and safety. J. Agric. Food Chem. **2012**, 60, 8535–8543.
8. Merilampi, S.; Bjorninen, T.; Ukkonen, L.; Ruuskanen, P.; Sydanheimo, L. Embedded wireless strain sensors based on printed RFID tag. Sens. Rev. **2011**, 31, 32–40.
9. Manzari, S.; Occhiuzzi, C.; Nawale, S.; Catini, A.; Di Natale, C.; Marrocco, G. Humidity Sensing by Polymer-Loaded UHF RFID Antennas. IEEE Sens. J. **2012**, 12, 2851–2858.
10. Bhattacharyya, R.; Floerkemeier, C.; Sarma, S. Low-Cost, Ubiquitous RFID-Tag-Antenna-Based Sensing. IEEE Proc. **2010**, 98, 1593–1600.

11. Jacquemod, G.; Nowak, M.; Colinet, E.; Delorme, N.; Conseil, F. Novel architecture and algorithm for remote interrogation of battery-free sensors. Sens. Actuators A Phys. **2010**, 160, 125–131.

12. Babar, A.A.; Manzari, S.; Sydanheimo, L.; Elsherbeni, A.Z.; Ukkonen, L. Passive UHF RFID Tag for Heat Sensing Applications. IEEE Trans. Antennas Propag. **2012**, 60, 4056–4064.

13. Cazeca, M.J.; Mead, J.; Chen, J.; Nagarajan, R. Passive wireless displacement sensor based on RFID technology. Sens. Actuators A Phys. **2013**, 190, 197–202.

14. Nilsson, H.E.; Unander, T.; Siden, J.; Andersson, H.; Manuilskiy, A.; Hummelgard, M.; Gulliksson, M. System Integration of Electronic Functions in Smart Packaging Applications. IEEE Trans. Compon. Pack. Manuf. Technol. **2012**, 2, 1723–1734.

15. Daliri, A.; Galehdar, A.; John, S.; Wang, C.H.; Rowe, W.S.T.; Ghorbani, K. Wireless strain measurement using circular microstrip patch antennas. Sens. Actuators A Phys. **2012**, 184, 86–92.

16. Li, X.Y.; Kan, E.C. A wireless low-range pressure sensor based on P(VDF-TrFE) piezoelectric resonance. Sens. Actuators A Phys. **2010**, 163, 457–463.

17. Yi, X.; Cho, C.; Cooper, J.; Wang, Y.; Tentzeris, M.M.; Leon, R.T. Passive wireless antenna sensor for strain and crack sensing—Electromagnetic modeling, simulation, and testing. Smart Mater. Struct. **2013**, 22, 085009.

18. Amin, Y.; Kanth, R.K.; Liljeberg, P.; Akram, A.; Chen, Q.; Zheng, L.-R.; Tenhunen, H. Printable RFID Antenna with Embedded Sensor and Calibration Functions. In Proceedings of the Progress in Electromagnetics Research Symposium, Stockholm, Sweden, 12–15 August 2013; pp. 567–570.

19. Gao, J.L.; Siden, J.; Nilsson, H.E.; Gulliksson, M. Printed Humidity Sensor with Memory Functionality for Passive RFID Tags. IEEE Sens. J. **2013**, 13, 1824–1834.

20. Abad, E.; Palacio, F.; Nuin, M.; de Zarate, A.G.; Juarros, A.; Gomez, J.M.; Marco, S. RFID smart tag for traceability and cold chain monitoring of foods: Demonstration in an intercontinental fresh fish logistic chain. J. Food Eng. **2009**, 93, 394–399.

21. Mattoli, V.; Mazzolai, B.; Mondini, A.; Zampolli, S.; Dario, P. Flexible tag datalogger for food logistics. Sens. Actuators A Phys. **2010**, 162, 316–323.

22. Steinberg, M.D.; Kassal, P.; Tkalcec, B.; Steinberg, M.I. Miniaturised wireless smart tag for optical chemical analysis applications. Talanta **2014**, 118, 375–381.

23. Oprea, A.; Courbat, J.; Barsan, N.; Briand, D.; de Rooij, N.F.; Weimar, U. Temperature, humidity and gas sensors integrated on plastic foil for low power applications. Sens. Actuators B Chem. **2009**, 140, 227–232.

24. Zampolli, S.; Elmi, I.; Cozzani, E.; Cardinali, G.C.; Scorzoni, A.; Cicioni, M.; Marco, S.; Palacio, F.; Gómez-Cama, J.M.; Sayhan, I.; et al. Ultra-low-power components for an RFID Tag with physical and chemical sensors. Microsyst. Technol. **2007**, 14, 581–588.

25. Kassal, P.; Steinberg, I.M.; Steinberg, M.D. Wireless smart tag with potentiometric input for ultra-low-power chemical sensing. Sens. Actuators B Chem. **2013**, 184, 254–259.

26. Unander, T.; Nilsson, H.E. Characterization of Printed Moisture Sensors in Packaging Surveillance Applications. IEEE Sens. J. **2009**, 9, 922–928.

27. Unander, T.; Siden, J.; Nilsson, H.E. Designing of RFID-Based Sensor Solution for Packaging Surveillance Applications. IEEE Sens. J. **2011**, 11, 3009–3018.

28. Fernandez, I.; Asensio, A.; Gutierrez, I.; Garcia, J.; Rebollo, I.; de No, J. Study of the communication distance of a MEMS Pressure Sensor Integrated in a RFID Passive Tag. Adv. Electr. Comput. Eng. **2012**, 12, 15–18.

29. Salmeron, J.F.; Molina-Lopez, F.; Rivadeneyra, A.; Quintero, A.V.; Capitan-Vallvey, L.F.; de Rooij, N.F.; Ozaez, J.B.; Briand, D.; Palma, A.J. Design and Development of Sensing RFID Tags on Flexible Foil Compatible with EPC Gen 2. IEEE Sens. J. **2014**, 14, 4361–4371.

30. Yang, L.; Rida, A.; Tentzeris, M.M. Design and Development of Radio Frequency Identification (RFID) and RFID-Enabled Sensors on Flexible Low Cost Substrates; Morgan & Claypool Publishers: San Rafael, CA, USA, 2009; Volume 1, pp. 1–89.

31. Vyas, R.; Lakafosis, V.; Rida, A.; Chaisilwattana, N.; Travis, S.; Pan, J.; Tentzeris, M.M. Paper-Based RFID-Enabled Wireless Platforms for Sensing Applications. IEEE Trans. Microw. Theory Tech. **2009**, 57, 1370–1382.

32. Kang, A.L.; Zhang, C.R.; Ji, X.J.; Han, T.; Li, R.S.; Li, X.W. SAW-RFID enabled temperature sensor. Sens. Actuators A Phys. **2013**, 201, 105–113.

33. Yin, J.; Yi, J.; Law, M.K.; Ling, Y.X.; Lee, M.C.; Ng, K.P.; Gao, B.; Luong, H.C.; Bermak, A.; Chan, M.; et al. A System-on-Chip EPC Gen-2 Passive UHF RFID Tag with Embedded Temperature Sensor. IEEE J. Solid State Circuit **2010**, 45, 2404–2420.

34. Catarinucci, L.; Colella, R.; Tarricone, L. Enhanced UHF RFID Sensor-Tag. IEEE Microw. Wirel. Compon. Lett. **2013**, 23, 49–51.

35. Broutas, P.; Contopanagos, H.; Kyriakis-Bitzaros, E.D.; Tsoukalas, D.; Chatzandroulis, S. A low power RF harvester for a smart passive sensor tag with integrated antenna. Sens. Actuators A Phys. **2012**, 176, 34–45.

36. Pletersek, A.; Sok, M.; Trontelj, J. Monitoring, control and diagnostics using RFID infrastructure. J. Med. Syst. **2012**, 36, 3733–3739.

37. Salmerón, J.F.; Rivadeneyra, A.; Agudo-Acemel, M.; Capitán-Vallvey, L.F.; Banqueri, J.; Carvajal, M.A.; Palma, A.J. Printed single-chip UHF passive radio frequency identification tags with sensing capability. Sens. Actuators A Phys. **2014**, 220, 281–289.

38. Paul Singh, S.; Burgess, G.; Singh, J. Measurement and analysis of the second-day air small and light-weight package shipping environment within federal express. Packag. Technol. Sci. **2004**, 17, 119–127.

39. Ostrem, F.E.; Godshall, W.D. An Assessment of the Common Carrier Shipping Environment. Available online: http://oai.dtic.mil/oai/oai?verb=getRecord&metadataPrefix=html&identifier=ADA071303 (accessed on 3 October 2015).

40. Campaign, U.F.D.; Maheras, S.J.; Lahti, E.A.; Ross, S.B. Transportation Shock and Vibration. Available online: http://www.pnnl.gov/main/publications/external/technical_reports/PNNL-22514.pdf (accessed on 2 September 2015).

41. Todd, B.; Phillips, M.; Schultz, S.M.; Hawkins, A.R.; Jensen, B.D. Low-Cost RFID Threshold Shock Sensors. IEEE Sens. J. **2009**, 9, 464–469.

42. AG, A. Demo Kit for the SL900A Smart EPC Sensor Tag IC. Available online: http://ams.com/eng/Support/Demoboards/UHF-RFID/UHF-Interface-and-Sensor-Tags/SL900A-Demo-Kit (accessed on 1 October 2015).

43. Rao, K.V.S.; Nikitin, P.V.; Lam, S.F. Antenna design for UHF RFID tags: A review and a practical application. IEEE Trans. Antennas Propag. **2005**, 53, 3870–3876.

44. Griffin, J.D.; Durgin, G.D. Complete link budgets for backscatter-radio and RFID systems. IEEE Antennas Propag. Mag. **2009**, 51, 11–25.

45. Salmerón, J.F.; Molina-Lopez, F.; Briand, D.; Ruan, J.J.; Rivadeneyra, A.; Carvajal, M.A.; Capitán-Vallvey, L.; de Rooij, N.F.; Palma, A.J. Properties and Printability of Inkjet and Screen-Printed Silver Patterns for RFID Antennas. J. Electron. Mater. **2014**, 43, 604–617.

9

CHAPTER

DESIGN OF AN ELLIPTIC CURVE CRYPTOGRAPHY PROCESSOR FOR RFID TAG CHIPS

Zilong Liu [1], Dongsheng Liu [1,*], Xuecheng Zou [1], Hui Lin [2] and Jian Cheng [1]

[1] *School of Optical and Electronic Information, Huazhong University of Science and Technology, Wuhan 430074, China*

[2] *School of Science, Wuhan University of Technology, Wuhan 430074, China*

ABSTRACT

Radio Frequency Identification (RFID) is an important technique for wireless sensor networks and the Internet of Things. Recently, considerable research has been performed in the combination of public key cryptography and RFID. In this paper, an efficient architecture of Elliptic Curve Cryptography (ECC) Processor for RFID tag chip is presented. We adopt a new inversion algorithm which requires fewer registers to store variables than the traditional schemes. A new method for coordinate swapping is proposed, which can reduce the complexity of the controller and shorten the time of iterative calculation effectively. A modified circular shift register architecture is presented in this paper, which is an effective way to reduce the area of register files. Clock gating and asynchronous counter are exploited to reduce the power consumption. The simulation and synthesis results show that the time needed for one elliptic curve scalar point multiplication over $GF(2^{163})$ is 176.7 K clock cycles and the gate area is 13.8 K with UMC 0.13 μm Complementary Metal Oxide Semiconductor (CMOS) technology. Moreover, the low power and low cost consumption make the Elliptic Curve Cryptography Processor (ECP) a prospective candidate for application in the RFID tag chip.

Keywords: wireless sensor network; RFID; Elliptic Curve Cryptography (ECC); processor

1. INTRODUCTION

Radio frequency identification (RFID) is an auto identification technology. Nowadays, it is widely used for identification control, chain management, wireless sensor networks (WSNs) and other applications. With the rapid development of the Internet of Things (ICT) and WSN, the demand on security-related RFID systems has grown fast [1]. These RFID applications require low-power and low-cost implementations of security mechanisms.

Recently, symmetric key cryptography, such as Advanced Encryption Standard (AES), has been suggested that it might not be preferable for RFID systems, since the number of RFID tags can be very large in WSNs and there is a potential risk involved in storing numerous symmetric keys. To satisfy security and system requirements, it is proved that a suitable public key cryptography scheme is necessary. Due to the fact that the traditional public key cryptography adds high overhead to the RFID tag chip, it has been considered to be unsuitable for a long time. For instance, passive RFID tags obtain energy from radio frequency signals, the power supply is limited. Therefore, these tags cannot utilize the energy-demanding cryptographic algorithms, such as the well-known RSA cryptography. Nevertheless, Elliptic Curve Cryptography (ECC), proposed by Koblitz [2], has been employed in many applications recently due to its numerous advantages over traditional public key cryptography schemes. The main advantage is that in utilizing the smaller key sizes, ECC can offer the similar security level as RSA [3]. For example, the security of 163-bit ECC is considered equivalent to 1024-bits RSA [4]. This feature makes it highly suited for implementation in RFID tag chips and being used in WSNs extensively.

The security of ECC is based on the difficulty of elliptic curve discrete logarithm problem (ECDLP) and the underlying operation in the elliptic curve cryptosystems is scalar point multiplication. The point multiplication can be performed by finite field arithmetic computations such as field addition, field multiplication, field squaring, and inversion. A number of hardware implementations for elliptic curve cryptography have been suggested in literatures, but only a few of them are aimed for RFID. Most implementations focus on speed and are based on the field-programmable gate array (FPGA) technology [5–9].

There are several implementations of scalar point multiplication in the literatures targeting RFID tag chips [10–13]. These implementations are different from coordinate systems (e.g., affine, projective, and mixed), basis (e.g., polynomial basis, normal basis), curves, architectures or algorithms. Most of these implementation efforts are concentrated on reducing the register number of elliptic curve cryptography processor without considering the practical applications, such as the transaction time. Generally, the total transaction time of RFID is required less than 300 ms [14]. However, most papers for RFID show that the time consumption to finish one point multiplication is over or close to 300 ms. Some papers just deliver a single ECC processor without involving the interfaces for other modules in a RFID tag chip, such as ROM, RAM and system controller. Low power consumption is an essential requirement for passive RFID tag chips. However, most related papers only focus on how to reduce the area of ECC processor for RFID tag chips. It has been ignored for a long time to reduce the power consumption with various low power techniques.

In this paper, we present an efficient implementation of Elliptic Curve Cryptography Processor (ECP) targeting RFID tag chips. First of all, the architecture of the RFID tag chip with ECP is described. Later on, we analyze different algorithms for scalar point multiplication and inversion, which are underlying operations in the elliptic curve cryptosystem. A new inversion algorithm is adopted and it needs fewer registers to store variables than the traditional methods. Then, we design a compact and efficient modular arithmetic logic unit (MALU). The modified circular shift register architecture is presented in this paper. We optimize the primary architecture and it shows an effective way to reduce the area of register files. In this paper, a new method of coordinates swapping is proposed which can reduce the complexity of the controller and shorten the time of iterative calculation effectively. In order to achieve low power consumption, clock gating and asynchronous counters are extensively used in our design. Finally, we perform an evaluation of our design and make comparisons with previous works.

The proposed design requires less clock cycles to finish one scalar point multiplication. The total time spent on scalar point multiplication is nearly 176.7 K clock cycles. The compact ECP needs only the area of 0.12 mm^2 with UMC 0.13 μm technology and the power consumption is 20.1 μW at the clock frequency of 847.5 KHz. All of these characteristics make the ECP design very attractive for RFID tag chips.

The rest of the paper is organized as follows: Section 2 presents the architecture of the RFID tag chip with our proposed ECP, and in Section 3, the ECC algorithms are analyzed and optimized. Section 4 presents the implementation design for the different modular arithmetic logic units. In Sections 5 and 6, the modified register array and the structure of ECP command controller are presented. In Section 7, the low power strategies are presented. Result analysis and comparison are carried out in Section 8. Finally, we conclude this paper in Section 9.

2. THE ARCHITECTURE OF THE RFID TAG CHIP WITH ECP

A typical ECP-embedded RFID tag chip can be divided into four parts [15], including Analog Front End (AFE), Random Number Generator (RNG), EEPROM and Digital Baseband Controller, as illustrated in Figure 1. AFE accomplishes the detailed functions of physical layer according to the RFID protocol, including carrier signal demodulation, modulation, power supply, clock generation, and reset signal generation. Random numbers generated from RNG will be used in elliptic curve digital signature algorithm (ECDSA). RNG can make sure the randomness property of each authentication so that the data in the authentication is unpredictable. EEPROM is used for storing private or public information, such as the private key, base point of elliptic curve (EC) and the EC equation parameters. Baseband Controller, utilizing the streamline bus structure, integrates the pre-processing circuit, RAM block, system controller, memory interface and ECP into one unit.

After the AFE demodulates a frame sent by the reader, then the pre-processing circuit will check the validation of the frame and extract the useful information from the frame. If the frame is legitimate, the RAM block will store the frame data into the memory arrays through the bus. When the phase of data receiving is over, the system controller will read the related information through the bus and load them into the ECP unit for further calculation. When the initialization of EC parameters has been fulfilled, the system controller will inform the ECP to start related calculation and wait for the signals from the ECP before entering the responding phase.

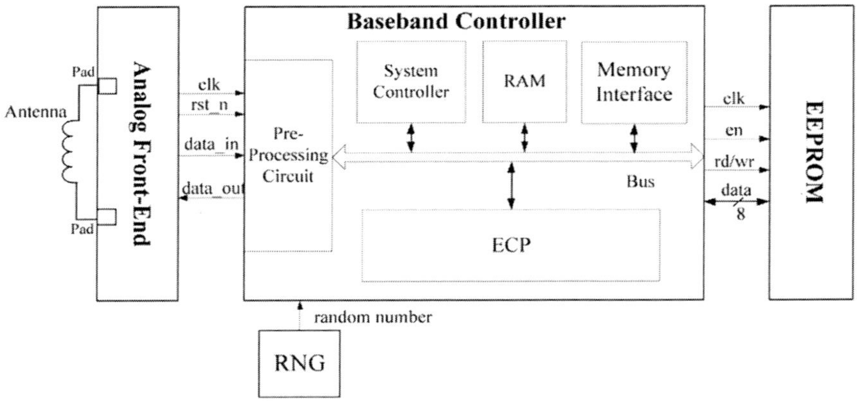

Figure 1. The structure of Radio Frequency Identification (RFID) tag with an Elliptic Curve Cryptography Processor (ECP).

In this paper, the structure of the ECP is suitable for Elliptic Curves (ECs) in binary extension field of GF(2^m), which means the value of m can be any legitimate value. However, for better discussion, m is set to 163, and the reduction polynomial is shown as follows:

$$F(x) = x^{163} + x^7 + x^6 + x^3 + 1 \quad (1)$$

As shown in Figure 2, we propose a structure of the ECP, composed of register array, MALU and ECP command controller. The ECP performs the scalar point multiplication, inversion and any other ECC calculations and returns the results of the calculations via the bus. In our implementation, the ECP loads the private key k, ECs parameter c and base point P(x,y) into register array from ROM and executes the scalar point multiplication or other operations. After finishing the calculation, it outputs the results for further use.

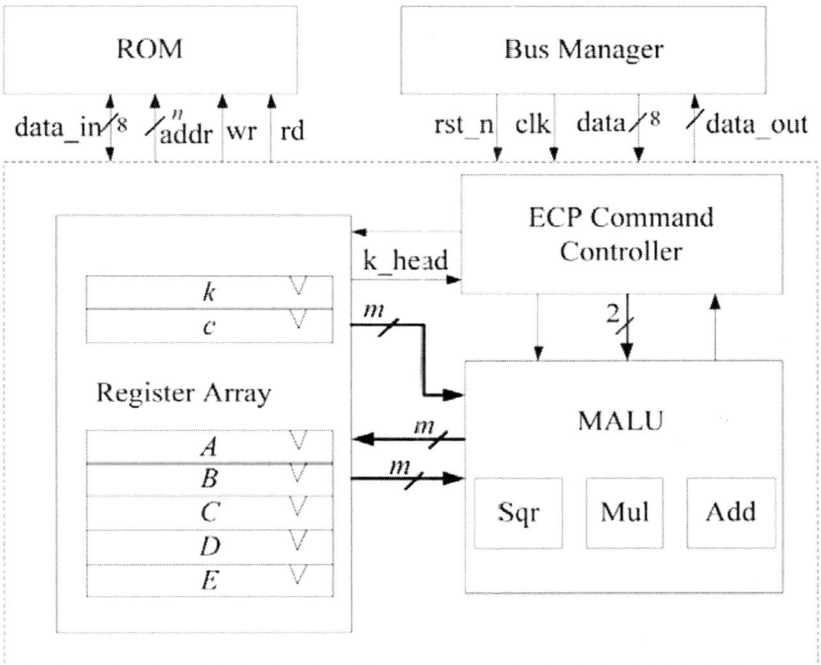

Figure 2. Structure of ECP.

3. SCALAR POINT MULTIPLICATION

3.1. Arithmetic Analysis

Scalar point multiplication, Q = k·P, is the underlying operation in the elliptic curve cryptosystems. P is the base point of ECs and k is a scalar used as private key. The resultant point Q will be used as a public key. According to ECDLP, if k is significantly large then it is very hard to retrieve k when the values of P and Q are given. The scalar point multiplication can be executed by point additions and point doublings, both of which involve many basic field arithmetic operations. In this paper, the EC is set as a generic Koblitz curve with the form of Equation (2), which is widely used in ECC, and the basic arithmetic operations are performed in the Galois field (GF). The GFs are either prime field GF(p) or extension binary field GF(2^m).

The GF(2^m) design is easier for hardware implementation and is adopted in this paper.

$$y^2 + xy = x^3 + ax^2 + b \tag{2}$$

There are two commonly used algorithms for scalar point multiplication, namely Binary Method [16] and Montgomery Ladder Method [17]. Binary Method is a basic scalar point multiplication method, also called double and add method, as shown in Algorithm 1. The scalar point multiplication iterates through every bit of k. In each iteration, the point doubling is performed. When the particular bit of k is one, the point addition is also performed. It means that the execution time of one scalar point multiplication is correlated to the hamming weight of the key k, and then the Simple Power Analysis (SPA) attacks become a threat to reveal the key value through recording power traces over time.

Algorithm 1. Binary Method for scalar point multiplication

Input : k *and* P, $k = (k_{m-1}, k_{m-2}, \ldots k_0)_2$, $k_i \in \{0,1\}$

Output : $Q = k \cdot P$

1.: *set* $Q \leftarrow 0$

2: ***for*** *i from* $m-1$ *to* 0 ***do***

3: $\quad Q \leftarrow 2Q$

4: $\quad$ ***if*** $k_i = 1$ ***then*** $Q \leftarrow Q + P$

5: *return* Q

Montgomery ladder method is one of the most commonly used algorithms to perform scalar point multiplication. The advantage of this method is that only the x-coordinate is used for point doubling and point addition in affine coordinates. Hence, the number of field operations and register variables can be reduced in each iteration. However, in affine coordinates, two inversion operations are needed in each iteration. As a result, at least 2 m field inversions should be performed to finish one scalar point multiplication, as shown in Algorithm 2, which leads to huge time consumption and is conflict with RFID real-time requirement. Therefore, we introduce the López-Dahab projective coordinate [18].

Algorithm 2. Montgomery Ladder Method in affine coordinates

Input: $P, k, where\ P = (x, y) \in GF(2^m), k = [k_{t-1} \cdots k_1, k_0] \in Z^+$

Output: $Q = kP = (x_1, y_1)$

$1: if\ k = 0\ or\ x = 0\ then\ return\ Q = (0, 0)\ and\ stop$

$2: set\ x_1 \leftarrow x, x_2 \leftarrow x^2 + b / x^2$

$3: for\ i = t - 2\ decre\ to\ 0\ do$

$$l = \frac{x_1}{(x_1 + x_2)}$$

$$if\ k_i = 1, set\ x_1 \leftarrow x + l^2 + l, x_2 \leftarrow x_2^2 + \frac{b}{x_2^2}$$

$$else\ set\ x_2 \leftarrow x + l^2 + l, x_1 \leftarrow x_1^2 + \frac{b}{x_1^2}$$

$4: set\ r_1 \leftarrow x_1 + x, r_2 \leftarrow x_2 + x$

$$5: set\ y_1 \leftarrow \frac{r_1(r_1 r_2 + x^2 + y)}{x} + y$$

$6: return\ Q(x_1, y_1)$

In projective coordinates, X-coordinate and Z-coordinate are used in each iteration, but the inversion operation is avoided. Only one inversion operation is performed in the conversion from projective coordinates to affine coordinates, as shown in Algorithm 3. It will dramatically reduce the time cost of the calculation. In the affine coordinates, because the inversion operation is performed in each iteration, 6 m-bit register variables are needed in the calculation of each iteration, which is the same number of m-bit registers as the method in the projective coordinates. Compared to Binary method, Montgomery ladder method performs point addition and point doubling in each iteration regardless of the value of k_i. It offers intrinsic protection against SPA and Timing Analysis (TA) [19].

In general, the scalar point multiplication can be divided into 4 basic computing elements in $GF(2^m)$, namely multiplication, addition, inversion and squaring. The inversion operation can be performed by multiplication and squaring. In this paper, we combine the adder, squarer and multiplier into one design unit called MALU.

Algorithm 3. Montgomery ladder algorithm in López-Dahab coordinates.

Input: $P, k, where\ P = (x, y) \in GF(2^m), k = [k_{t-1} \cdots k_1, k_0] \in Z^+$

Output: $Q = kP = (x_1, y_1)$

$1: if\ k = 0\ or\ x = 0\ then\ return\ Q = (0,0)\ and\ stop$

$2: set\ X_1 \leftarrow x, Z_1 \leftarrow 1, X_2 \leftarrow x^4 + b, Z_2 \leftarrow x^2$

$3: for\ i = t-2\ decreto\ 0\ do$

$if\ k_i = 1, set\ (X_1, Z_1) = Madd(X_1, Z_1, X_2, Z_2)$

$(X_2, Z_2) = Mdouble(X_2, Z_2)$

$else\ set\ (X_2, Z_2) = Madd(X_2, Z_2, X_1, Z_1)$

$(X_1, Z_1) = Mdouble(X_1, Z_1)$

$4: set\ Q(x_1, y_1) = Mxy(X_1, Z_1, X_2, Z_2)$

$5: return\ Q(x_1, y_1)$

$Note:$

$$Madd = \begin{cases} X_{add} = x \times Z_{add} + (X_1 \times Z_2) \times (X_2 \times Z_1) \\ Z_{add} = (X_1 \times Z_2 + X_2 \times Z_1)^2 \end{cases}$$

$$Mdouble = \begin{cases} X_{double} = X_2^4 + b \times Z_2^4 \\ Z_{double} = (X_2 \times Z_2)^2 \end{cases}$$

$$Mxy = \begin{cases} x_1 = \dfrac{X_1}{Z_1} \\ y_1 = y + \dfrac{X_1 \times (X_1 \times X_2 + x \times X_1 \times Z_2 + x \times X_2 \times Z_1 + y \times Z_1 \times Z_2)}{x \times Z_2 \times Z_1^2} \end{cases}$$

3.2. Inversion Algorithm

In the end of Algorithm 3, Mxy(X_1, Z_1, X_2, Z_2) means the conversion from projective coordinates to affine coordinates. It requires only one inversion operation in Algorithm 3. As inversion is a complicated operation, the Fermat's Little Theorem provides a simple method to perform inversion as follows [20]:

$$A^{-1} = A^{2^m - 2} = A^{2(1+2+\ldots+2^{m-2})} \tag{3}$$

According to Equation (3), in the binary extension field of GF(2^{163}), a total of 161 multiplications and 162 squarings are needed in one inversion operation. The previous Itoh-Tsujii (IT) algorithm is an effective scheme to reduce the number of multiplication operation in one inversion [21]. Algorithm 4 shows the inversion operation with the IT method. It is based on the following decomposition:

$$1+2+2^2+...2^{161}=(2\times(2\times(1+2)(1+2^2)+1)\times(1+2^5)\times(1+2^{10})\times(1+2^{20})\times(1+2^{40})+1)\times(1+2^{81}) \quad (4)$$

Algorithm 4. Inversion with IT method over GF(2^{163}).

Input : $A \in GF(2^{163})$

Output : $T = A^{-1} = A^{2^m-2}$ where $m = 163$

$1: B \leftarrow A^2$	$2: T \leftarrow B \times A,$
$3: B \leftarrow T^{2^2}$	$4: T \leftarrow B \times T,$
$5: B \leftarrow T^2$	$6: T \leftarrow B \times A,$
$7: B \leftarrow T^{2^5}$	$8: T \leftarrow B \times T,$
$9: B \leftarrow T^{2^{10}}$	$10: B \leftarrow B \times T,$
$11: B \leftarrow T^{2^{20}}$	$12: T \leftarrow B \times T,$
$13: B \leftarrow T^{2^{40}}$	$14: T \leftarrow B \times T,$
$15: B \leftarrow T^2$	$16: T \leftarrow B \times A,$
$17: B \leftarrow T^{2^{81}}$	$18: T \leftarrow B \times T,$
$19: T \leftarrow T^2$	

Recently, a new Dimitrov-Järvinen (DJ) algorithm for inversion was introduced [22]. The new DJ method is based on the ingenious decomposition as follows [23]:

$$1+2+2^2+\ldots 2^{161}=(1+2+2^2)\times(1+2^3+2^6)\times(1+2^9+2^{18})\times(1+2^{27}+2^{54})\times(1+2^{81}) \quad (5)$$

From Equations (4) and (5), in the binary extension field of GF(2^{163}), IT method and DJ method both require 9 multiplications and 162 squarings. In the implementation of these two methods, IT method needs three 163-bit register variables, but DJ method only needs two 163-bits register variables. Our design has implemented the DJ method in projective coordinate system with only one inversion operation. The inversion algorithm with DJ method is shown in Algorithm 5.

Algorithm 5. Inversion with the Dimitrov-Järvinen (DJ) method over GF(2^{163}).

Input : $A \in GF(2^{163})$

Output : $T = A^{-1} = A^{2^m-2}$ where $m = 163$

$1: A \leftarrow A^2$	$2: T \leftarrow A^2,$
$3: A \leftarrow A \times T$	$4: T \leftarrow T^2,$
$5: A \leftarrow A \times T$	$6: T \leftarrow A^{2^3},$
$7: A \leftarrow A \times T$	$8: T \leftarrow T^{2^3},$
$9: A \leftarrow A \times T$	$10: T \leftarrow A^{2^9},$
$11: A \leftarrow A \times T$	$12: T \leftarrow T^{2^9},$
$13: A \leftarrow A \times T$	$14: T \leftarrow A^{2^{27}},$
$15: A \leftarrow A \times T$	$16: T \leftarrow T^{2^{27}},$
$17: A \leftarrow A \times T$	$18: T \leftarrow A^{2^{81}},$
$19: T \leftarrow A \times T$	

4. MODULAR ARITHMETIC LOGIC UNIT (MALU)

The MALU shown in Figure 3 performs the GF(2^m) field operations, namely the multiplication, addition, squaring and inversion. The inversion operation can be performed by multiplications and squarings. The MALU contains 5 units, an adder unit, multiplier unit, squarer unit, controller unit and register T. The MALU is designed to execute 4 basic commands, namely ADD, SQR, MUL and INV. The operands for these commands are stored in the register array. Opcode is used to select which operation to be performed. When the signal of malu_en is valid and the MALU gets the input operands through the bus of A_bus and B_bus, the MALU will make proper operation according to the value of Opcode.

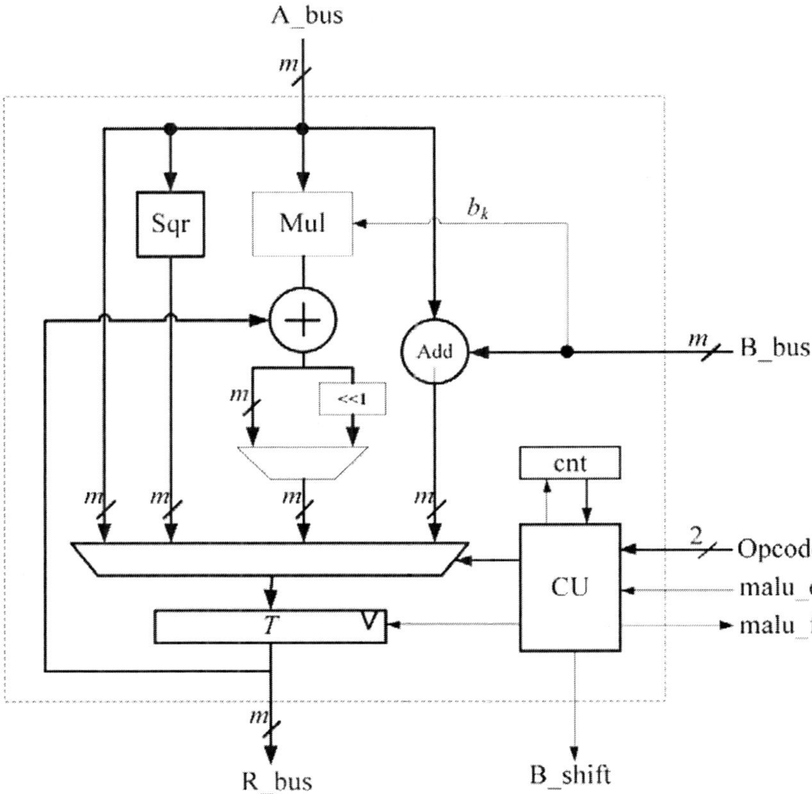

Figure 3. Structure of the modular arithmetic logic unit (MALU).

For addition, the field addition operation adds A and B, provided by A_bus and B_bus. It can be simply obtained by a bit-wise XOR operation. For squaring, the field squaring operation needs only one operand provided by A_bus. Addition and squaring operations can be efficiently performed in the MALU with a latency of one clock cycle.

Multiplication is more complex. By using the bit-serial multiplier, one operand is provided by A_bus. The multiplier requires the most significant bit (MSB) b_k of B_bus for the multiplicand. The signal B_shift is used for multiplication and is provided to Reg_B in the register array. In the field of $GF(2^m)$, when the multiplication is performed, B_shift will be valid for m clock cycles, and the register Reg_B will shift to the left m times. The counter in the MALU is used to count the number of clock cycle for multiplication. It will take m clock cycles to finish one multiplication operation with the bit-serial multiplier.

After each command is executed, the MALU will provide a signal called malu_fin to inform up-level ECP for further computing. The register T serves as a medium for restoring the results of 4 basic commands and is responsible for exchanging data with register Reg_A through R_bus.

4.1. Adder Unit

For two elements, $A=(a_{m-1},a_{m-2},\ldots,a_0)\in GF(2^m)$ and $B=(b_{m-1},b_{m-2},\ldots,b_0)\in GF(2^m)$, field addition in the binary extension field of $GF(2^m)$ can be simply obtained by a bit-wise XOR addition operation as $A+B=\sum_{i=0}^{m-1}(a_i \oplus b_i)x^i$ Therefore, the Adder Unit is implemented in our design using 163 XOR gates with one clock cycle output latency.

4.2. Squarer Unit

We represent $A(x)\in GF(2^m)$ in polynomial basis as follows:

$$A(x)=\sum_{i=0}^{m-1} a_i x^i,\ a_i \in GF(2) \tag{6}$$

Squaring in $GF(2^m)$ can be calculated as follows:

$$C(x)=A^2(x)=a_{m-1}x^{2m-2}+\cdots+a_2x^4+a_1x^2+a_0 \bmod F(x) \tag{7}$$

In the field of GF(2^{163}), m=163, the field arithmetic is implemented as polynomial arithmetic modulo F(x). Using the equivalence in Equation (8), it can reduce the double-sized result with the reduction polynomial.

$$x^{163} = x^7 + x^6 + x^3 + 1 \tag{8}$$

From Equations (7) and (8), we can get the results as follows:

$$\begin{aligned}
c_{162}x^{162} &= a_{81}x^{162} + a_{159}x^{162} + a_{161}x^{162} \\
c_{161}x^{161} &= a_{159}x^{161} + a_{162}x^{161} \\
&\cdots \\
c_4x^4 &= a_2x^4 + a_{82}x^4 + a_{160}x^4 \\
c_3x^3 &= a_{83}x^3 + a_{160}x^3 + a_{161}x^3 \\
c_2x^2 &= a_1x^2 + a_{161}x^2 \\
c_1x^1 &= a_{82}x^1 + a_{160}x^1 + a_{162}x^1 \\
c_0 &= a_0 + a_{160}
\end{aligned}$$

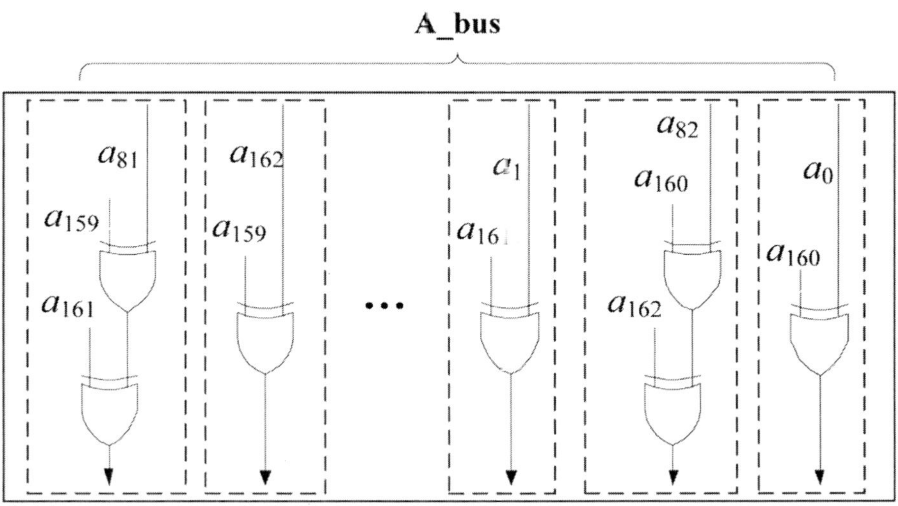

Figure 4. Squaring Circuit in GF(2^{163}).

Therefore, without huge area requirements, squaring in GF(2^m) can be efficiently implemented as a combinational logic circuit to output the result in one clock cycle. There are total 247 XOR gates in the squaring circuit. The operand comes from A_bus and the result can be saved in register T. The squarer unit is shown in Figure 4.

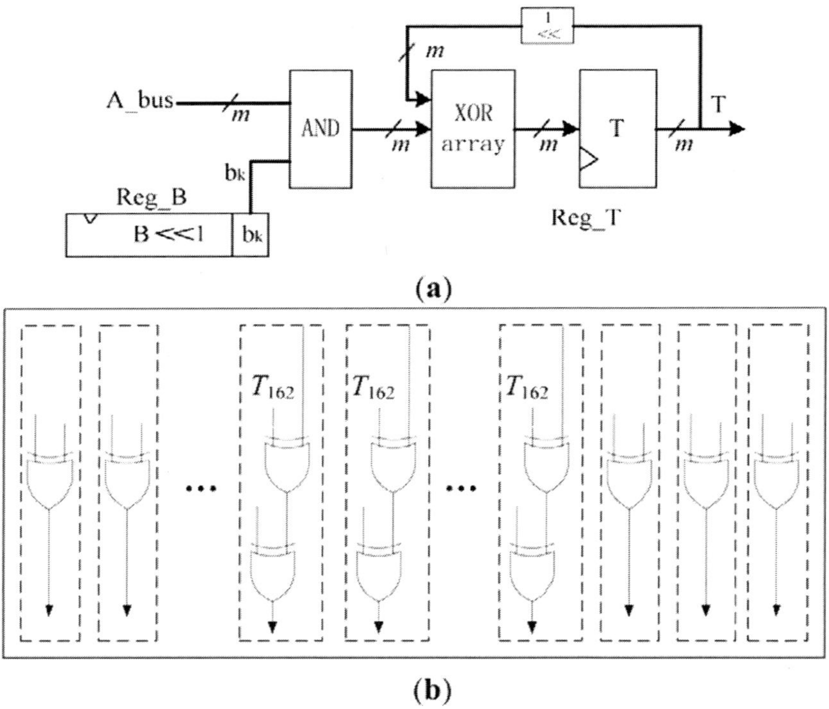

Figure 5. (**a**) The architecture of bit-serial multiplier Circuit in GF(2^{163}); (**b**) The XOR array for multiplier.

4.3. Multiplier Unit

The multiplier unit is the biggest component in the MALU. As the low-power and low-cost requirements of RFID, it is necessary to optimize the design of multiplier unit. Bit-serial multipliers are proved to be the most efficient scheme that can reduce area consumption and maintain good performance [11]. It will take m clock cycles to finish one multiplication operation. We implement the Bit-serial multipliers as Algorithm 6, and the structure is shown in Figure 5. The operand A can be enabled onto the A_bus, directly from the Register Array. The individual bit of b_i comes from the MSB of B_bus which is directly from a 163-bits cyclic shift-register Reg_B in the Register Array. The XOR array is based on the reduction polynomial and is composed of 166 XOR gates. Therefore, there are total 163 AND gates, 166 XOR gates and a 163-bits shifter-register T in the bit-serial multiplier circuit. The register T is also used to store the results of addition and squaring.

Algorithm 6. Bit-serial multiplication in the field of GF(2^{163}).

Input : $A(x)=\sum_{i=0}^{m-1} a_i x^i,\ B(x)=\sum_{i=0}^{m-1} b_i x^i,\ where\ a_i, b_i \in \{0,1\}$

Output : $T \equiv A \cdot B \bmod F(x)=\sum_{i=0}^{m-1} c_i x^i,\ where\ c_i \in \{0,1\}$

$1: T \leftarrow 0$

$2:$ **for** $i=m-1\ down\ to\ 0$ **do**

$$T \leftarrow b_i \cdot A(x) + T \cdot x + t_{m-1} \cdot F(x)$$

$3:$ **Return** T

5. Register Array

The architecture of the register array file is shown in Figure 6. This circular shift register architecture has been proved to be an effective method to reduce the area consumption of register files [10]. We modify the primary architecture and make it fit for our whole design. In [10], the calculation that converting projective coordinates to affine coordinates is not involved and the register k is set in their controller module. Therefore, there are only six registers needed in their register file. In our register file, there are seven registers including register k. Register k and register c are the constant registers. Reg_A, Reg_B, Reg_C, Reg_D, Reg_E are used to store temporary variables of the calculation. Register c has an 8-bit I/O through which data coming from external ROM or RAM can be loaded and stored. These external data can be private key k, ECs parameter c, or the coordinates of base point P(x,y). When the ECP needs some constant parameters for calculation, register c will load those parameters. After the private key k is loaded into register c from the external ROM, register k will load the key value from register c immediately. After finishing the data preparation of part 2 in Algorithm 3, the register k can be shifted by one bit to the left according to the iteration calculation of part 3 in Algorithm 3 and provides the k_i value for ECP command controller.

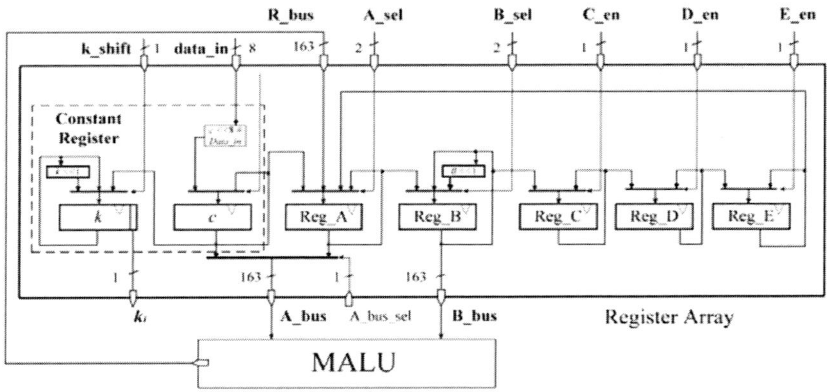

Figure 6. Register array file architecture.

The whole register array is a circular shift register file. Each register is independently controlled by ECP command controller for efficient management. Furthermore, Reg_B is a circular shift register which can be shifted by one bit to the left. The data in Reg_A and Reg_B will input to MALU through A_bus and B_bus for the calculation based on opcode. Reg_C, Reg_D, Reg_E can only be updated by the preceding register, as the original scheme. To reduce the power consumption, clock gating, as a low power strategy, is applied to the circular shift register. Compared to the primary architecture, we integrate the register of private key k and the register of constant c into the register array. It makes our ECP become more compact and efficient.

6. ECP COMMAND CONTROLLER

The ECP command controller, as shown in Figure 7, executes data preparation, point addition, point doubling, and the Mxy according to Algorithm 3. The data preparation is carried out simply by transferring (x,y) into (X_1, Z_1, X_2, Z_2), from the affine coordinates to the López-Dahab projective coordinates. The Mxy part involves the final calculations for converting projective coordinates to affine coordinates. The routines implemented in the ECP command controller are collected in Figure 8.

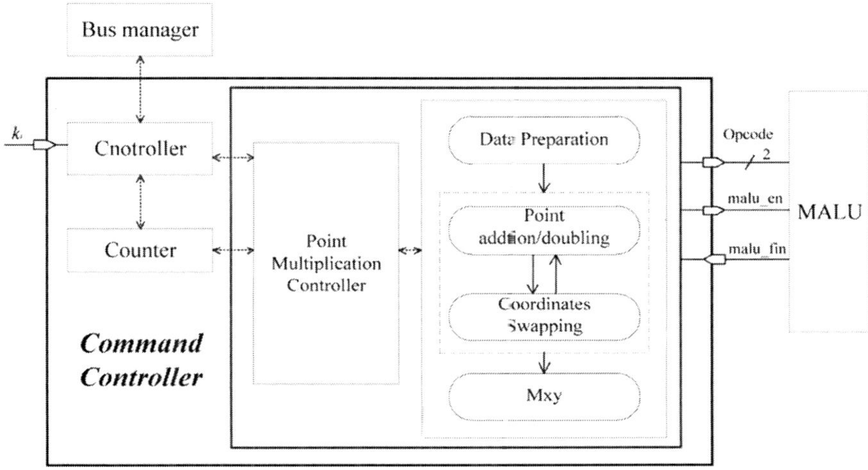

Figure 7. ECP Command Controller.

Data Preparation

1: $X_1 \leftarrow x$
2: $Z_1 \leftarrow 1$
3: $Z_2 \leftarrow x^2$
4: $X_2 \leftarrow Z_2^2$
5: $A \leftarrow c^2$
6: $X_2 \leftarrow X_2 + A$

Point addition :

$(X_1, Z_1) = \text{Madd}(X_1, Z_1, X_2, Z_2)$
1: $X_1 \leftarrow X_1 \times Z_2$
2: $Z_1 \leftarrow Z_1 \times X_2$
3: $A \leftarrow X_1 \times Z_1$
4: $Z_1 \leftarrow Z_1 + X_1$
5: $Z_1 \leftarrow Z_1^2$
6: $X_1 \leftarrow Z_1 \times x$
7: $X_1 \leftarrow X_1 + A$

Point double :

$(X_2, Z_2) = \text{Mdouble}(X_2, Z_2)$
8: $X_2 \leftarrow X_2^2$
9: $Z_2 \leftarrow Z_2^2$
10: $A \leftarrow Z_2 \times c$
11: $Z_2 \leftarrow Z_2 \times X_2$
12: $A \leftarrow A^2$
13: $X_2 \leftarrow X_2^2$
14: $X_2 \leftarrow X_2 + A$

Coordinates Swapping (if k_i == 0)

1: $A \leftarrow X_1$
2: $X_1 \leftarrow X_2$
3: $X_2 \leftarrow A$
4: $A \leftarrow Z_1$
5: $Z_1 \leftarrow Z_2$
6: $Z_2 \leftarrow A$

Projective → Affine coordinate

$Q(x_1, y_1) = Mxy(X_1, Z_1, X_2, Z_2)$
1: **if**($Z_1 == 0$)
    return $Q(0,0)$
2: **if**($Z_2 == 0$)
    return $Q(x, x+y)$
3: $A \leftarrow Z_1 \times Z_2$
4: $Z_1 \leftarrow Z_1 \times x$
5: $Z_1 \leftarrow Z_1 + X_1$
6: $Z_2 \leftarrow Z_2 \times x$
7: $X_1 \leftarrow Z_2 \times X_1$
8: $Z_2 \leftarrow Z_2 + X_2$
9: $Z_2 \leftarrow Z_2 \times Z_1$
10: $B \leftarrow x^2$
11: $B \leftarrow B + y$
12: $B \leftarrow B \times A$
13: $B \leftarrow B + Z_2$
14: $A \leftarrow A \times x$
15: $A \leftarrow A^{-1}$
16: $B \leftarrow B \times A$
17: $X_2 \leftarrow X_1 \times A$
18: $Z_2 \leftarrow X_2 + x$
19: $Z_2 \leftarrow Z_2 \times B$
20: $Z_2 \leftarrow Z_2 + y$
21: $x_1 \leftarrow X_2$
22: $y_1 \leftarrow Z_2$

Figure 8. Routines implemented in the controller.

According to part 3 of Algorithm 3, the ECP command controller takes in the most significant bit (MSB) of the shift register k from register array. No matter if k_i = 0 or k_i = 1, a point addition and a point doubling are performed in one iterative calculation. Figure 8 shows that five registers are needed to store temporary variables in the calculation of point addition and point doubling. The base point coordinate x and EC parameter c will be loaded into the constant register c from the external ROM when they are needed.

In the iterative calculation of the part 3 of Algorithm 3, when $k_i = 1$, point addition and point doubling are performed as follows:

$$A \leftarrow X_1 Z_2 Z_1 X_2 , Z_1 \leftarrow (X_1 Z_2 + Z_1 X_2)^2 , X_1 \leftarrow x Z_1 + A \quad (9)$$

$$A \leftarrow c^2 Z_2^4 , Z_2 \leftarrow X_2^2 Z_2^2 , X_2 \leftarrow X_2^4 + A \quad (10)$$

when $k_i = 1$, at the beginning of point addition, X_1, Z_1, X_2, Z_2 will be stored in Reg_A, Reg_B, Reg_C Reg_D respectively. The result of point addition will be stored in X_1, Z_1. In other words, the value of Reg_A and Reg_B will be updated with the new value of X_1 and Z_1. Reg_C and Reg_D will keep the previous value of X_2 and Z_2. As soon as the calculation of point addition is finished, the calculation of point doubling will begin. The inputs to this operation are X_2, Z_2, and EC parameter c. By the way of circular shift, register array will provide the right operand to MALU, and the previous value of X_1, Z_1 will be kept in the circular shift. The outputs to point doubling are saved in X_2, Z_2. Therefore, the values in Reg_A, Reg_B, Reg_C and Reg_D will be updated with the new value of X_1, Z_1, X_2, Z_2, respectively, after the calculation of one point addition and point doubling. Four registers are utilized to store the value of X_1, Z_1, X_2, Z_2, and another one is used to store the value of A.

When $k_i = 0$, point addition and point doubling are performed as follows:

$$A \leftarrow X_1 Z_2 Z_1 X_2 , Z_2 \leftarrow (X_1 Z_2 + Z_1 X_2)^2 , X_2 \leftarrow x Z_2 + A \quad (11)$$

$$A \leftarrow c^2 Z_1^4 , Z_1 \leftarrow X_1^2 Z_1^2 , X_1 \leftarrow X_1^4 + A \quad (12)$$

According to Equations (9) and (11), when k_i changes from 1 to 0, the terms $X_1Z_2Z_1X_2$ and $X_1Z_2 + Z_1X_2$ will remain unchanged. When $k_i = 0$, before the calculation of point addition and point doubling, X_1 and Z_1 will be swapped with X_2 and Z_2. Therefore, X_1, Z_1, X_2, Z_2 will be stored in Reg_C, Reg_D, Reg_A, Reg_B respectively. The outputs to point addition are saved in X_2, Z_2 (Reg_A, Reg_B) and the outputs to point doubling are saved in X_1, Z_1 (Reg_C, Reg_D). As the updated result value of X_1, Z_1, X_2, Z_2 are stored in Reg_C, Reg_D, Reg_A, Reg_B respectively, a coordinate swapping is still needed to swap X_1, Z_1 with X_2, Z_2, in the end of point doubling. Therefore, the final results of X_1, Z_1, X_2, Z_2 are stored in Reg_A, Reg_B, Reg_C, Reg_D, respectively.

Consequently, the point addition and point doubling can be repeated in the iterative calculation in the part 3 of Algorithm 3 without the involvement of k_i bits. However, when k_i = 0, it needs extra coordinates swapping operations. By the way of coordinates swapping, it is easy to implement the controller and decrease the complexity. In addition, it can shorten the time of iterative calculation effectively.

As Figure 8 shows, in data preparation, it needs three squaring and one field addition. Six multiplications, five squaring and three field additions are performed in the calculation of one point addition and point doubling. In the phase of converting projective coordinates to affine coordinates, including an inversion operation, there are total of 19 multiplications, 163 squaring and 6 field additions. Assuming H(k) = 163, the hamming weight of private k, point multiplication will take approximately 163.9 K clock cycles without clock cycles of controlling. When the base point coordinate P(x, y) or EC parameter c is utilized as one operand, the constant register requires a read operation, and the operation will take 21 clock cycles. The final simulation result shows that point multiplication will take 176.7 K clock cycles based on our whole design scheme.

7. LOW POWER STRATEGIES

As the energy provided for passive RFID tag comes from radio frequency signals, low power consumption is an essential requirement for passive RFID tag chip. The operation distance of RFID systems depends on the maximum of the dynamic power of tag chip. Clock-gating and asynchronous counters are adopted in the design of our ECC processor to minimize power consumption.

Clock gating is a popular technique used in many synchronous circuits for reducing dynamic power dissipation [24]. Clock gating saves power by adding more logic to a circuit to prune the clock tree. Pruning the clock disables portions of the circuitry so that the flip-flops in them do not have to switch states. The used clock gating cells come from UMC standard cell library in 0.13 μm CMOS technology.

Many counters are used in the design of ECC processor. Using synchronous counter might bring in large power consumption, since every bit of the registers would be triggered on each of the clock edges. For asynchronous counter, as shown in Figure 9, only the first flip-flop would be triggered by clk. The subsequent flip-flops are triggered by the former flip-flops. Therefore, the unnecessary switches in registers can be minimized as well as reducing the dynamic power. As the ECC processor for RFID tag works at a low clock frequency, there is no need to use synchronous counters. Compared to synchronous counter, asynchronous counters with more than 4-bit flip-flops structure can reduce the power consumption by 50% at least [25].

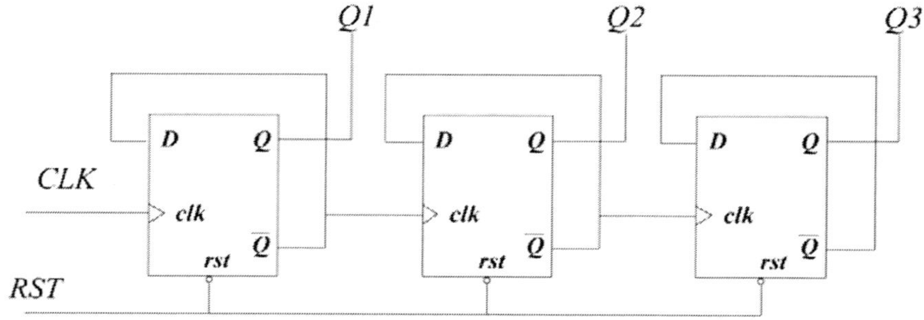

Figure 9. Asynchronous counter.

8. RESULT AND COMPARISON

8.1. Result Analysis

The proposed design has been conducted in Verilog HDL. We synthesized our processor using a low leakage power library of UMC 0.13 μm (fsc0l_d_generic_core_tt1p2v25c.db). For synthesis, we used the EDA tool of Design Complier. Gate equivalent (GE) can provide meaningful comparisons of area among different technologies. The area of two inputs NAND is 5.12 μm^2 in fsc0l lib. Table 1 shows the gate count of every component of our ECP. The controller and the MALU have the area of 918 GE and 3591 GE respectively. The register array (9342 GE) dominates the gate area and occupies 67% area of the whole design.

Table 1. Gate area of individual processor components.

	Ecp_con	MALU	Reg_array
Cell Area (μm^2)	4702.7	18385.9	47831.1
Area (Gates)	918	3591	9342

Power consumption is one of the most important factors for passive RFID tag chip. We used EDA tools of VCS and Power Compiler to get the estimation of average power at the gate level. First, the Value Change Dump (VCD) files were generated in VCS. Then, VCD files were translated to Switching Activity Interchange Format (SAIF) files, which were used in Power Complier to get average power consumptions. The carrier frequency of ISO/IEC 14443 standard is 13.56 MHz [26], so the source clock frequency provided by AFE is 13.56 MHz.When the ECP is applied to the RFID tag chip, the ECP will be under the control of the baseband. The work clock of ECP comes from the baseband. The source clock may be divided by 2^n, and the divided clock will be provided for the ECP, where n = 0, 1, 2, 3, 4, 5, 6. Figure 10 and Table 2 show the ECP's power consumption, time and energy for one scalar point multiplication at different work frequencies and the power consumption can be effectively reduced by decreasing the operation frequency. Because of the adoption of low power strategies, such as clock gating and asynchronous counter, it can save about 15% power consumption at the same work clock frequency. In order to compare the frequency power benefits, Figure 11 shows the normalized time, power and energy for one scalar point multiplication under different work frequencies.

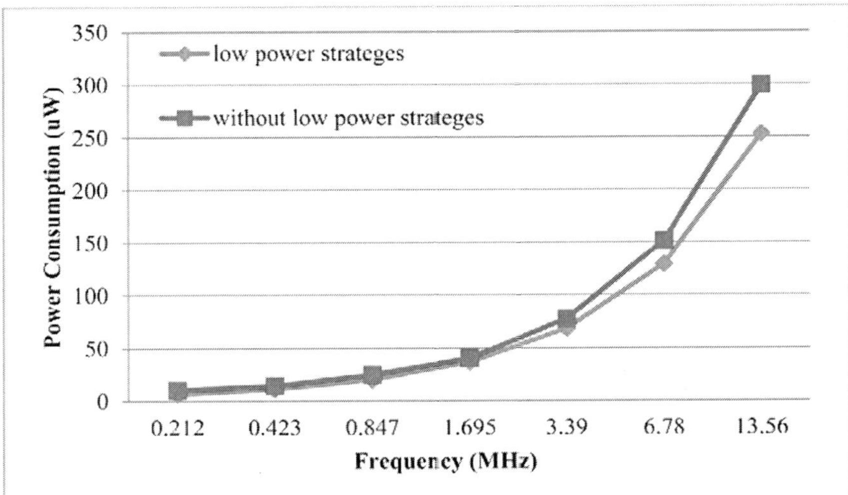

Figure 10. Power consumption under different work frequencies.

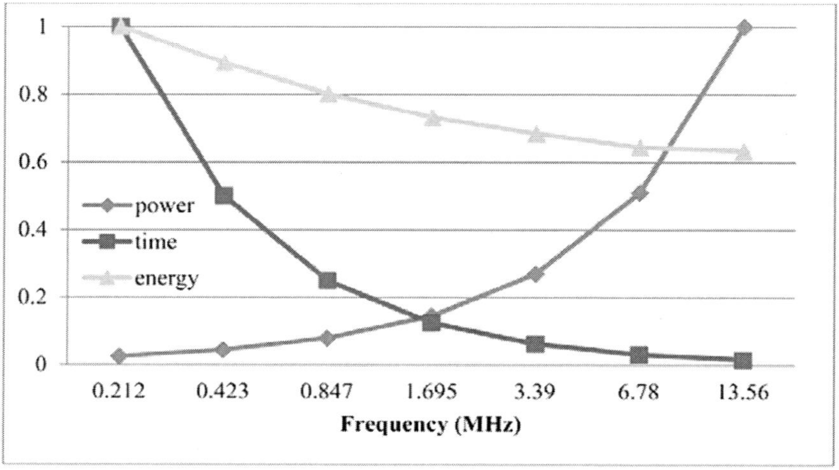

Figure 11. Normalized time, power and energy for one scalar point multiplication under different work frequencies.

Table 2. Power, time and energy consumptions under different work frequencies.

Freq (KHz)	13,560	6780	3390	1695	847.5	423	212
Power (μW)	253	129	68.6	36.7	20.1	11.2	6.27
Time (ms)	13.1	26.1	52.2	104.5	208.5	417.7	833.5
Energy (μJ)	3.31	3.37	3.58	3.83	4.19	4.68	5.23

The simulation result of a certain scalar point multiplication demonstrates that the time needed for one calculation is 176.7 K clock cycles. Taking the ISO/IEC 14443 standard as an example [26], the AFE will extract a clock with the frequency of 13.56 MHz, so the total time consumption is nearly 13.1 ms.

As the ISO 14443 standard specified, the frame waiting time (FWT) defines the maximum time for a RFID tag to start its response frame after the end of a reader frame, as shown in Figure 12. The FWT is calculated by the following formula:

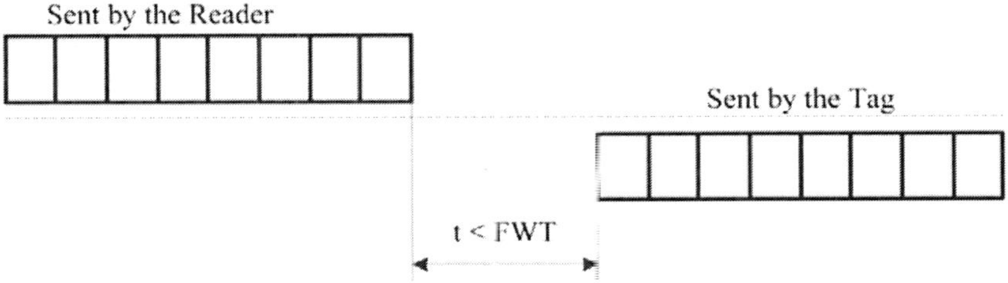

Figure 12. Frame waiting time.

$$FWT = (256 \times 16 / fc) \times 2^{FWI} \tag{13}$$

where fc is the carrier frequency and where FWI is the frame waiting time integer. FWI is coded in the range from 0 to 14. The default value of FWI is 4, which means that FWT = 4.8 ms by default. However, when the tag needs more time to process the information it receives, such as the ECC operations, it can impose the reader to increase the FWI up to 6, which corresponds to FWT = 19.3 ms. In fact, the total transaction time of most RFID applications is required less than 300 ms. For example, a validation process at a gate should take less than 300 ms usually [27]. The Hong Kong mass transit smart card takes about 140–300 ms for one transaction [28]. The smaller FWT can reduce the customers' waiting time effectively. Our design requires 13.1 ms for one ECC operation at 13.56 MHz. Therefore, it is compatible with the ISO 14443 standard and most RFID applications, such as mass transit smart card.

Both the power consumption and the calculation time prove that our proposed ECP can meet the demands of real-time and resource-constraint for the applications of RFID tag chips. The layout of the ECP is shown in Figure 13. The largest module is the register array, followed by the MALU module, and the ECP command controller is the smallest module. The overall area of the ECP processor is 349 μm × 358 μm.

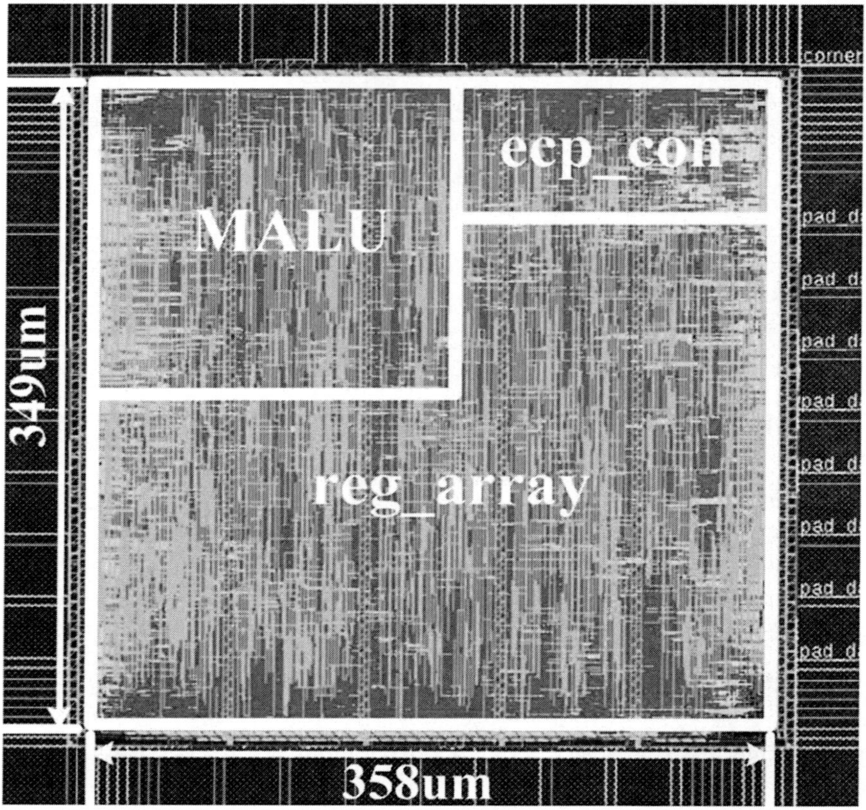

Figure 13. Final Layout Result.

8.2. COMPARISONS

Table 3 shows the comparison with the related works. The ECP can work at different clock frequencies according to the real-time and power requirements of different RFID systems. The number of clock cycles for one scalar point multiplication is 176.7 K and it is much less than the design of [10–13].

Table 3. Comparison with related works.

Ref.	Freq. (KHz)	Power (μW)	Cycles	Time (ms)	ECP Area (gates)	Core Size (μm^2)	Energy * (μJ)	Technology
This Work	13.560	253	176.7 K	13.1	13.8 K	124,942	3.3	UMC 0.13 μm
	1.690	36.7		104.5			3.8	
	847.5	20.1		208.5			4.2	
	423	11.2		417.7			4.7	
[10]	1,130	36.63	275.8 K	244.08	12.5 K	N.A.	8.9	UMC 0.13 μm
	590	21.55	144.8 K	245.49	14.1 K		5.3	
	411	15.75	101.2 K	246.19	14.7 K		3.9	
	323	12.08	78.5 K	243.17	15.4 K		2.9	
[11]	13.560	N.A.	376.9 K	31.8	15 K	N.A.	N.A.	AMI 0.35 μm
[12]	400	7.3	219.1 K	547.87	11.7 K	N.A.	3.9	UMC 0.13 μm
[13]	847.5	83	297 K	350.4	13.2 K	219,897	29.1	UMC 0.18 μm
	106	10.8		2801.9			30.2	
	106	54.7		2801.9		N.A.	153.2	AMS 0.35 μm

* Energy consumption for one scalar point multiplication.

The MALU in the design of [10] has an 8-bit adder and an 8-bit multiplier. Therefore, the squaring operation uses the same logic as the multiplication, hence, each squaring requires m clock cycles. In our design, there is a specialized squarer unit and the squaring operation requires only one clock cycle. Moreover, the design of [10] assumes that the coordinate conversion to the affine coordinate system and the calculation of Y-coordinate value are performed on a reader or back-end system, hence, the inversion operation has not been implemented in [10]. If the inversion is performed in the design of [10], another 163-bit register should be added. The area of one bit register composed of D-type flip-flops is about 8 GE in UMC 0.13 μm technology. In fact, the coordinate conversion is necessary in a complete ECP of the RFID tag chip for encryption and decryption applications, such as ECDSA [29]. Due to the adoption of the DJ method, one 163-bit register is saved in the inversion operation of our design. As a result, the number of 163-bit register is the same as [10], but the coordinate conversion is implemented in our scheme. The design of [10] can acquire the trades-offs between the gate area and the number of clock cycles depending on the digit sizes, but the time cost for one scalar point multiplication is about 240 ms, it can hardly satisfy the transaction time requirements of some RFID systems.

The literature [12] has adopted the same method and structure as [10] does. The ECs parameters are fixed that can be very helpful to reduce the area, but it is not compatible in different RFID systems. In our design, the register c of register array can load in different ECs parameters. Moreover, the EC in [12] is a kind of binary Edwards curve, which is a special elliptic curve. The binary Edwards curves have not been recommended by NIST standards [30] and are not widely used in cryptosystems. The Koblitz curve is widely used and is adopted in our design.

In [11], the ECP has been implemented in an affine coordinate system, with the disadvantage that it requires one inversion to be computed in each iteration. As a result, it needs much more clock cycles than our work for one scalar point multiplication. Moreover, the ECs parameter c in [11] is fixed. Hence, the design of [11] cannot be used in other Koblitz curves when the parameter c is changed. In [11], it does not achieve a much better result than ours or other works. Finally, in [13], the scalar point multiplication is done in 297 K clock cycles and the area is 13.2 K GE. The gate area of our design is slightly larger than [13], but the number of clock cycles is much less than [13].

Due to the adoption of clock gating and asynchronous counter, the power consumption is smaller than all the other designs at the same clock frequency. With the power and the time to finish one scalar point multiplication, we can get the energy consumption demonstrating that our design can satisfy the low-power requirement of RFID tag chips. The low number of clock cycles needed for one scalar point multiplication makes it suitable for real-time applications of RFID tag chips. Although the gate area of our design is slightly larger than [12,13], our design achieves a good trade-off between the real-time requirement and the constrained resource.

9. CONCLUSIONS

A design of Elliptic Curve Cryptography processor based on Koblitz curves for the RFID tag chip is presented in this paper. We employ Montgomery Ladder algorithm for scalar point multiplication which is the underlying operation in the elliptic curve cryptography. A new DJ method is exploited for inversion operation and it requires fewer registers to store variables than the traditional methods. A preferable MALU for our ECP design is presented in detail. In the iterative calculation of point addition and point doubling, a new coordinates swapping method is proposed. It can reduce the complexity of the ECP command controller and shorten the time of iterative calculation effectively. The modified circular shift register file is introduced to reduce the complexity of system. Some low power strategies, such as clock gating and asynchronous counter are adopted, saving about 15% power consumption.

Our design was synthesized with UMC 0.13 μm CMOS technology at different frequencies. Compared to other reported results, our architecture acquires good trade-offs of clock cycle number, gate area, power consumption and energy. On the aspects of power consumption and time cost, our design shows better performance. According to the implementation results, the ECP area is 0.12 mm^2, and the power consumption can be reduced to 20.1 μW at the clock frequency of 847.5 KHz, which can meet the demands of real-time, low-power, and resource-constraints for the applications of RFID tag chips and WSNs.

ACKNOWLEDGMENTS

This work is supported by the National Natural Science Foundation of China (Grant No.61306038 and No.61376031), ChenGuang Project of science and technology for Wuhan youths (No.2013072304010829) and the Fundamental Research Funds for the Central Universities (HUST: 2013TS050).

AUTHOR CONTRIBUTIONS

All authors have contributed to the presented work. Zilong Liu, Dongsheng Liu and Xuecheng Zou proposed the architecture and scheme. Zilong Liu and Jian Cheng conducted the implementation and wrote the manuscript. Hui Lin has contributed to algorithm analysis. All authors have read and revised the manuscript.

CONFLICTS OF INTEREST

The authors declare no conflict of interest.

REFERENCES

1. Atzori, L.; Morabito, G. The internet of things: A survey. Comput. Netw. **2010**, 54, 2787–2805. [Google Scholar]
2. Koblitz, N. Elliptic curve cryptosystems. Math. Comput. **1987**. [Google Scholar]
3. Koblitz, N.; Menezes, A.; Vanstone, S. The state of elliptic curve cryptography. Des. Codes Cryptogr. **2000**, 19, 173–193. [Google Scholar]
4. National Security Agency. The Case for Elliptic Curve Cryptography. Available online: http://www.nsa.gov/business/programs/elliptic_curve.shtml (accessed on 15 September 2014).

5. Mahdizadeh, H.; Masoumi, M. Novel architecture for efficient FPGA implementation of elliptic curve cryptographic Processor over $GF(2^{163})$. IEEE Trans. Very Large Scale Integr. **2013**, 21, 2330–2333. [Google Scholar]
6. Gustavo, D.; Jeanm, P.D.; Jose, L.I. Efficient elliptic curve point multiplication using digit-serial binary field operation. IEEE Trans. Ind. Electron. **2013**, 60, 217–225. [Google Scholar]
7. Järvinen, K. Optimized FPGA-based elliptic curve cryptography processor for high-speed applications. Integr. VLSI J. **2011**, 44, 270–279. [Google Scholar]
8. Dimitrov, V.S.; Järvinen, K.U.; Jacobson, M.J.; Chan, W.F.; Huang, Z. Provably sublinear point multiplication on Koblitz curves and its hardware implementation. IEEE Trans. Comput. **2008**, 57, 1469–1481. [Google Scholar]
9. Azarderakhsh, R.; Masoleh, A.R. High-Performance implementation of point multiplication on Koblitz curves. IEEE Trans. Circuits Syst. II Exp. Briefs. **2013**, 60, 41–45. [Google Scholar]
10. Lee, Y.K.; Sakiyama, K.; Batina, L.; Verbauwhede, I. Elliptic-curve-based security processor for RFID. IEEE Trans. Comput. **2008**, 57, 1514–1527. [Google Scholar]
11. Kumar, S.; Paar, C. Are standards compliant elliptic curve cryptosystems feasible on RFID? Proceedings of Workshop on RFID Security, Graz, Austria, 12–14 July 2006; pp. 12–14.
12. Kocabas, U.; Fan, J.; Verbauwhede, I. Implementation of binary Edwards curves for very-constrained devices. Proceedings of 2010 the 21st International Conference on Application-Specific Systems Architectures and Processor (ASAP), Rennes, France, 7–9 July 2010; pp. 185–191.
13. Hein, D.; Wolkerstorfer, J.; Felber, N. ECC is Ready for RFID-A Proof in Silicon. In Selected Areas in Cryptography; Roberto, M.A., Liam, K., Eds.; Springer: Heidelberg, Germany, 2009; Volume 5381, pp. 401–413. [Google Scholar]
14. Wang, D.M.; Ding, Y.Y.; Zhang, J.; Hu, J.G.; Tan, H.Z. Area-Efficient and ultra-low-power architecture of RSA Processor for RFID. Electron. Lett. **2012**, 48, 1185–1187. [Google Scholar]
15. Nainan, S.; Romin, P.; Tanvi, S. RFID Technology Based Attendance Management System. Int. J. Comput. Sci. Issue **2013**, 10, 516–521. [Google Scholar]
16. Menezes, A.; Oorschot, A.V.; Vanstone, S. Handbook of Applied Cryptography; CRC Press Inc.: Boca Raton, FL, USA, 1997. [Google Scholar]
17. Montgomery, P. Speeding the pollard and elliptic curve method of factorization. Math. Comput. **1987**, 48, 243–264. [Google Scholar]

18. López, J.; Dahab, R. Fast Multiplication on elliptic curves over GF(2^m) without precomputation. In Cryptographic Hardware and Embedded Systems; Cetin, K.K., Christof, P., Eds.; Springer: Heidelberg, Germany, 1999; Volume 1717, pp. 316–327. [Google Scholar]
19. Joye, M.; Yen, S.M. The Montgomery Powering Ladder. In Cryptographic Hardware and Embedded Systems; Burton, S.K., Cetin, K.K., Eds.; Springer: Heidelberg, Germany, 2002; Volume 2523, pp. 291–302. [Google Scholar]
20. Burton, D.M. Elementary Number Theory; Tata McGraw-Hill Education: Noida, India, 2006. [Google Scholar]
21. Itoh, T.; Tsujii, S. A fast algorithm for computing multiplicative inverses in GF(2^m) using normal bases. Inf. Comput. **1988**, 78, 171–177. [Google Scholar]
22. Dimitrov, V.; Järvinen, K. Another look at inversions over binary fields. Proceedings of 2013 21st IEEE Symposium on Computer Arithmetic (ARITH), Austin, TX, USA, 7–10 April 2013; pp. 211–218.
23. Reza, A.; Kimmo, U.J.; Mechran, M.K. Efficient algorithm and architecture for elliptic curve cryptography for extremely constrained secure applications. IEEE Trans. Circuits Syst. I Regul. Pap. **2014**, 61, 1144–1155. [Google Scholar]
24. Wu, Q.; Pedram, M.; Wu, X. Clock-Gating and its application to low power design of sequential circuits. IEEE Trans. Circuits Syst. I Fundam. Theory Appl. **2000**, 47, 415–420. [Google Scholar]
25. Wei, D.; Zhang, C.; Cui, Y.; Chen, H.; Wang, Z. Design of a low-cost low-power baseband-processor for UHF RFID tag with asynchronous design technique. Proceedings of the 2012 IEEE International Symposium on Circuits and Systems (ISCAS), Seoul, Korea, 20–23 May 15.
26. Identification Cards—Contactless Integrated Circuit(s) Cards—Proximity Cards, Part 1: Physical Characteristics, ISO/IEC 14443-1, December 1999.
27. Ferreira, J.; Filipe, P.; Cunha, G.; Silva, J. Cloud terminals for ticketing systems. Proceedings of the Fifth International Conferences on Advanced Service Computing, Valencia, Spain, 27–31 May 2013; pp. 19–25.
28. Markantonakis, K.; Mayes, K.; Sauveron, D.; Askoxylakis, I.G. Overview of security threats for smart cards in the public transport industry. Proceedings of the 2008 IEEE International Conference on e-Business Engineering, Xi'an, China, 22–24 October 2008; pp. 506–513.

29. Johnson, D.; Menezes, A.; Vanstone, S. The elliptic curve digital signature algorithm (ECDSA). Int. J. Inf. Secur. **2001**, 1, 36–63. [Google Scholar]
30. National Institute of Standards and Technology. Recommended Elliptic Curves for Federal Government Use. Available online: http://csrc.nist.gov/groups/ST/toolkit/documents/dss/NISTReCur.pdf (accessed on 15 September 2014).

10

CHAPTER

DESIGN AND IMPLEMENTATION OF A RF POWERING CIRCUIT FOR RFID TAGS OR OTHER BATTERYLESS EMBEDDED DEVICES

Dongsheng Liu, Rencai Wang *, Ke Yao, Xuecheng Zou and Liang Guo

School of Optical and Electronic Information, Huazhong University of Science & Technology, Wuhan 430074, China

ABSTRACT

A RF powering circuit used in radio-frequency identification (RFID) tags and other batteryless embedded devices is presented in this paper. The RF powering circuit harvests energy from electromagnetic waves and converts the RF energy to a stable voltage source. Analysis of a NMOS gate-cross connected bridge rectifier is conducted to demonstrate relationship between device sizes and power conversion efficiency (PCE) of the rectifier. A rectifier with 38.54% PCE under normal working conditions is designed. Moreover, a stable voltage regulator with a temperature and voltage optimizing strategy including adoption of a combination resistor is developed, which is able to accommodate a large input range of 4 V to 12 V and be immune to temperature variations. Latch-up prevention and noise isolation methods in layout design are also presented. Designed with the

HJTC 0.25 μm process, this regulator achieves 0.04 mV/°C temperature rejection ratio (TRR) and 2.5 mV/V voltage rejection ratio (VRR). The RF powering circuit is also fabricated in the HJTC 0.25 μm process. The area of the RF powering circuit is 0.23 × 0.24 mm^2. The RF powering circuit is successfully integrated with ISO/IEC 15693-compatible and ISO/IEC 14443-compatible RFID tag chips.

Keywords: batteryless embedded devices; wireless sensor network; RFID tag chip; rectifier; regulator

1. INTRODUCTION

The applications of RFID technologies, including supply chain management, access control to buildings, public transportation, airport baggage handling, express parcel delivery and logistics, have increased rapidly in recent years [1–3]. Owing to its wireless and far field measurement capabilities, the combination of RFID technology and wireless sensing networks has also found a large number of applications such as environmental monitoring and healthcare devices [4–6]. RFID tags are often classified as passive or active. The passive tag is powered by an electromagnetic RF (radio frequency) wave transmitted by a reader, while the active tag is powered by a battery. The passive tag has advantages of low cost and long life, thus enable sensing nodes in a sensing system to be placed in positions that are not easily accessible [7,8].

However, passive RFID tags, and other batteryless embedded devices such as contactless smart cards, wireless sensor nodes and implantable biomedical devices, have to work at a considerably long distance from the transmitter [9,10]. Moreover, power converted from the electromagnetic RF wave to the tag should be sufficient to supply device in its operating distance, especially when a power hungry sensor is integrated. As the RF energy received by the tag (transponder) or other batteryless embedded devices decreases rapidly with distance, the induced voltage across the tag antenna is often variable in big interval and sensitive to temperature and process variation [11,12].

Thus, key challenges to implement a passive tag are harvesting energy with high efficiency and generating a stable voltage source for a digital baseband controller, the memory of the tag and its functional parts. This voltage source must be immune to process, input voltage and environmental temperature changes. Meanwhile, in order to meet the requirements for a more complicated RFID tag such as integrating sensors and CPU in a tag chips, the RF power circuit should be capable of handling heavy current loads [13,14].

To solve those problems, many RF power circuit design techniques for passive RFID tags have been reported to improve the power conversion efficiency and stability of the regulator power supply [14–17].

A low power RF harvester for a smart passive sensor tag used in structural health monitoring applications was proposed in [16]. The RF harvester harvests energy from the RF field and provides a stable and controllable power supply. Measurements of this design indicate full operation of the sensor tag for time intervals of 20 ms for up to 4 m distances with a tag reader power of 200 mW. However, this RF harvester is a chipless design while tags that utilize an application-specific chip are an undeniable trend for passive tag applications.

For specific tag chip designs, a low power CMOS voltage regulator for a wireless blood pressure biosensor is presented in [8]. A low voltage regulator, as part of the associated electronics of a blood pressure bio-medical implanted device, provides the power supply for the system. The line regulation of this regulator is 39 mV/V and its total power consumption is 1.2 mW. Since the body temperature is kept stable at approximately 37 °C, the regulator of this design is not optimized for temperature changes.

Moreover, to improve the stability of the power supply, a RF powering circuit which is immune to variations of temperature and voltage is proposed in [14]. A ripple stabilizer and a temperature stabilizer have been integrated to maintain a stable reference voltage that is insensitive to temperature variation and voltage ripples. Its line regulation is 12 mV/V and temperature rejection ratio is 0.2 mV/°C. However, the two huge resistors used for meeting the requirement of the low static current occupy a large area.

As for achieving high PCE, a RF to DC CMOS rectifier with high efficiency over a wide input power range for RFID applications is implemented in [17]. The rectifier is based on a novel active load circuit which adjusts the output current as a function of the incoming RF power. This design allows maximizing both the efficiency and sensitivity of the circuit and achieves a highest PCE of 45%.

In this paper, a RF powering circuit which is immune to process, voltage and temperature changes with high PCE is designed for a HF passive RFID tag chip. Compared with previous works, the input voltage of the regulator ranges from 4 V to 12 V which is much larger than that reported in the previous works, while a stable output voltage is also maintained against temperature and voltage variations. Furthermore, a rather high PCE is achieved in this work, which means this RF powering circuit can provide more power to the tag and hence enable further identification distance and it can be applied in more complicated passive tags.

The paper is organized as follows: Sections 2 and 3 describe the design and implementation of the RF powering circuit. The simulation and measurement results are presented in Section 4 and followed by conclusions in Section 5.

2. ARCHITECTURE OF THE RF POWERING CIRCUIT

A passive RFID tag generally consists of an analog front end (AFE), a digital baseband and an EEPROM, where the EEPROM can be divided into a logic control part and a memory part. If the tag is integrated with other specific functions such as sensing or monitoring, a functional block can be added to the tag system. In this design, the digital baseband and the logic control part of EEPROM are realized with standard cells, while the analog design is responsible for the AFE and the memory part of EEPROM.

As shown in Figure 1, the RF powering circuit comprises four parts. First, the energy harvesting circuit, consisting of an inductor and a resonance capacitor, harvests energy from an electromagnetic field through inductive coupling and resonance. A magnetic coil is used as antenna in this design. Then, the rectifier circuit, which consists of a bridge rectifier and a capacitor, converts the input RF signal from electromagnetic field into a DC signal with ripples. Consisting of a bias circuit and regulator V_{CC}/V_{DD},the regulator is designed to convert the DC signal with ripples into a stable power supply. The output voltage of the regulator has much smaller ripples and can drive several loads. Finally, an over-voltage protection circuit is presented to limit voltage of antennas by bleeding spare energy in antennas into the ground and this ensures that the whole chip has a stable and reliable operation.

The regulator V_{CC} drives the AFE and the memory part of the EEPROM. The average load current of the regulator V_{CC} is 100 μA, in which 80 μA is for the memory array and 20 μA is for the AFE. The regulator V_{DD} drives the digital baseband controller and the logic control of the EEPROM. The load current of the regulator V_{DD} is 420 μA, where 20 μA is for the EEPROM logic control, and 400 μA is for baseband controller and other functional blocks.

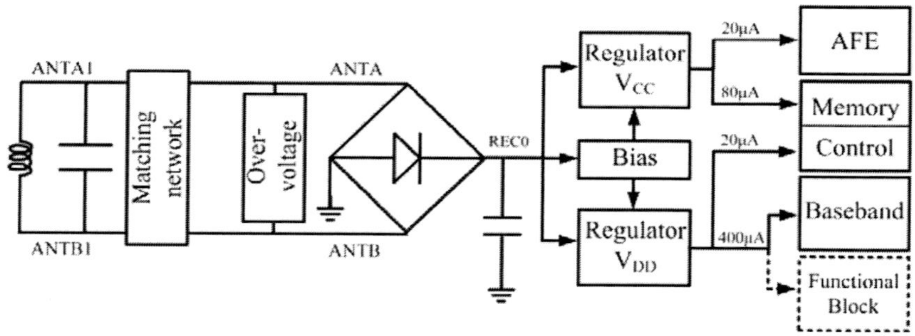

Figure 1. Structure of the RF powering circuit.

In our design, the digital and analog circuits are driven by V_{DD} and V_{CC} respectively, which makes it flexible to design the driving capability for different circuits, therefore meeting the requirements of high power consumption chips such as digital baseband controllers based on FSM, security chips, CPUs and chips integrated with sensors. Meanwhile, the huge noise of the digital circuit can be isolated by driving the digital and analog part separately. Thus the cost of noise restriction in the AFE can be minimized in this way.

3. DESIGN AND IMPLEMENTATION OF RF POWERING CIRCUIT

3.1. Rectifier Circuit

As shown in Figure 2, a NMOS gate cross-connected bridge rectifier structure, which is usually used in RF transponders for its high PCE, is adopted as the rectifier circuit.

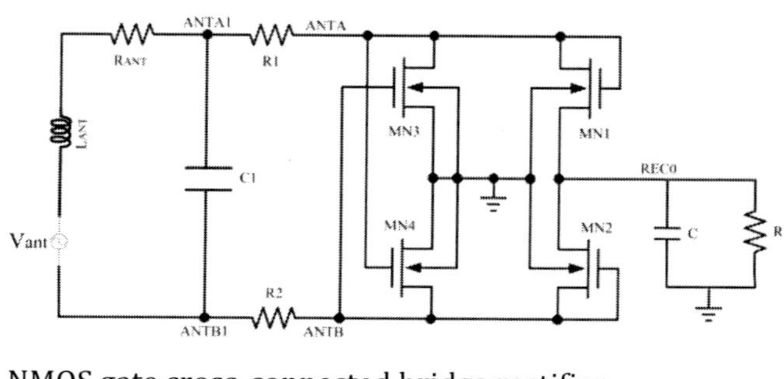

Figure 2. NMOS gate cross-connected bridge rectifier.

In Figure 2, the gates of NMOS transistors MN3, MN4 are cross connected to the antennas and MN1, MN2 are connected as diodes. The antenna can be equivalent to an AC voltage source V_{ant}, an inductor L_{ANT} and the parasitic resistor R_{ANT}. Resistors R1, R2 form the matching circuit, which matches the impedances of the antenna and the rectifier, thus L_{ANT} and C1 resonate and the maximum voltage values of ANTA and ANTB will be obtained. The impedance of R1 and R2 is obtained according to the input impedance of the cascading circuit to the resistors R1 and R2.

The equivalent circuit of the NMOS gate cross-connected bridge rectifier is shown in Figure 3a. Resistors R1, R2 form the matching circuit, and R_L is the equivalent input impedance of the cascading circuit to the resistors R1 and R2, V_{ANT} is the input voltage of the rectifier. Resistors R1, R2, and R_L are connected to C1 in parallel. They can be equivalently transformed to be connected in series as shown in Figure 3b. In Figure 3b, R_{LS} is obtained after the parallel to serial impedance conversion of resistors R1, R2, and R_L. An approximative value of R_{LS} is calculated by the equation below:

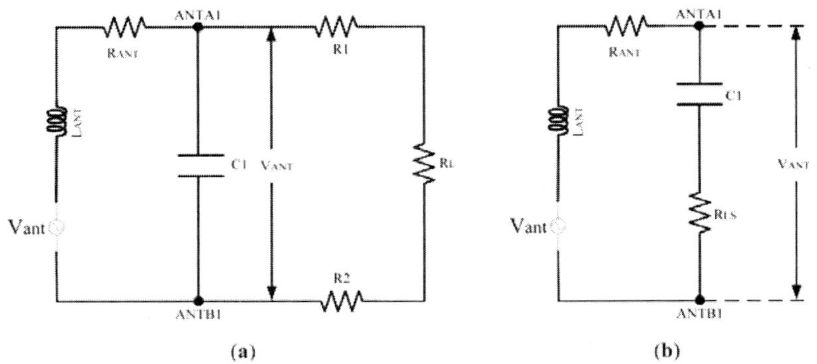

Figure 3. (**a**) Equivalent circuit of the NMOS gate cross-connected bridge rectifier; (**b**) Equivalent circuit of (a) after parallel to serial impedance conversion.

$$R_{LS} = \frac{1}{(R_L + R_1 + R_2)\cdot(\omega C1)^2} \tag{1}$$

According to the series resonance theory, the quality factor Q of the circuit in Figure 3b can be obtained by:

$$Q = \frac{1/\omega C1}{1/(R_1 + R_2 + R_L)\cdot(\omega C1)^2 + R_{ANT}} \quad (2)$$

When the value of Q is greater, the input voltage V_{ANT} will be higher. According to Equation (2), to make Q bigger, the values of R1 and R2 should be big as well. As the value of R_{ANT} is 20 Ω, the value of C1 is 16.5 pF, so the values of resistors R1 and R2 are both designed to be 15.7 kΩ to make the previous factor of the denominator of Equation (2) infinitesimally smaller than the following one.

When ANTA is high and ANTB is low, MN1, MN4 are on and MN2, MN3 are off. The capacitor is charged through path ANTA-MN1-C-GND-MN4-ANTB. When ANTB is high and ANTA is low, the capacitor will be charged through path ANTB-MN2-C-GND-MN3-ANTA.

In this circuit, the threshold voltage loss is as low as a V_{TH}, which is the threshold voltage of NMOS. As MN3 and MN4 do not produce any V_{TH} loss, the threshold voltage loss is produced by MN1 and MN2. It can be calculated that the minimum operation input level is slightly higher than V_{TH} because of the substrate effect. Thus, the minimum operation input level is efficiently reduced in this design. According to the preceding analysis, the antenna voltage is similar to a sinusoidal signal, and the output voltage level is [18]:

$$V_{RECO} = V_{ANT} / \sqrt{2} - V_{TH} \quad (3)$$

where V_{RECO} is the output voltage of the rectifier, V_{ANT} is the peak amplitude of the antennas voltage, V_{TH} is the threshold voltage of NMOS. As $V_{SB} = V_{RECO}$, V_{SB} is the source-bulk voltage of MN1 and MN2, according to the bulk effect model of NMOS, there is:

$$V_{TH} = V_{TH0} + \gamma(\sqrt{|2\phi_F + V_{RECO}|} - \sqrt{2\phi_F}) \quad (4)$$

where V_{TH0} is the intrinsic threshold voltage, γ is the body effect coefficient, ϕ_F is the Fermi potential. According to Equations (3) and (4), V_{RECO} increases with respect to V_{ANT}. As the electromagnetic field strength changes from 150 mA/m to 5 A/m, which is defined in ISO/IEC 15693 standard and related ISO/IEC 10373-7 test standard, the output voltage of the rectifier varies a lot from 4 V to 12 V according to the simulation. As a rectifier converts a RF signal of antenna to a DC voltage, higher PCE of the rectifier means more energy is converted to the tag. PCE is calculated by the equation below:

$$PCE = \frac{P_{out}}{P_{in}} \tag{5}$$

PCE can be also calculated by PCE = $P_{out}/(P_{out} + P_{mos})$, where P_{mos} is the power consumption of the transistors of the rectifier. P_{mos} can be calculated by, $P_{mos} = P_{dynamic} + P_{static}$, where $P_{dynamic}$ is the power consumption of the transistors while the transistors are on, and P_{static} is the power consumption of the transistors while they are off. $P_{dynamic} = \overline{V_{on} \cdot I_{on}}$, V_{on} is the forward voltage drop of the transistors, and I_{on} is the forward current of the transistors. $P_{static} = \overline{V_{reverse} \cdot I_{reverse}}$, $V_{reverse}$ is the reverse voltage of the transistors, and $I_{reverse}$ is the reverse leakage current of the transistors. The turn-on voltage of the rectifier determines the minimum input-voltage of the rectifier, higher turn-on voltage means higher Von, and greater reverse leakage current means greater P_{static}, so while the turn-on voltage is higher, and the reverse leakage current is greater, then the input power will be greater.

According to the aforementioned analysis of resistors R1 and R2, the resistors make PCE higher. As proper R1 and R2 bring a biggest Q of the circuit, so the value of V_{ANT} will be maximum when L_{ANT} and C1 resonate. Bigger V_{ANT} means greater P_{out}, then the value of PCE will be higher.

As shown in Table 1, PCE is greatly influenced by the size of the MOS device. According to the ISO/IEC 15693 standard and the related ISO/IEC 10373-7 test standard, the input voltage of the rectifier is always higher than 4 V, so the V_{ANT} is set to be 6 V as a normal value in this simulation. The load current of this RF powering circuit is less than 600 μA, and the output power of the antenna is normally less than 5 mW, so the R_{load} is set to be 6 kΩ. In this table, MOS devices have the same width/length (represented as W/L afterwards) ratio and M is the multiplier of a MOSFET. Greater M value means bigger size of a MOSFET. Bigger size of a rectifier MOSFET brings smaller channel conduction resistor, thus the output voltage of the rectifier is increased and higher PCE is achieved. However, bigger size of a MOSFET increases Ids of the MOS device, hence it consumes more power. As a result, when M > 20, the increase of PCE becomes slower than that when M is smaller. Tradeoffs between PCE and power consumption should be considered when designing the rectifier.

Table 1. Simulation results of NMOS gate cross-connected bridge rectifier (R_{load} = 6 kΩ, $\mathbf{V_{ANT}}$ = 6 V).

M	VHD/V	Pin/mW	Pout/mW	PCE
5	2.64	2.60	1.16	44.60%
10	3.01	3.00	1.51	50.23%
20	3.30	3.30	1.83	55.30%
30	3.46	3.53	2.00	56.60%
40	3.55	3.68	2.10	57.05%
50	3.62	3.78	2.18	57.70%

3.2. Linear Regulator

The V_{CC} voltage regulator is shown in Figure 4. By using a combination resistor and a calculation strategy for the device parameters, an optimization for regulator performance related to voltage and temperature is achieved in this work. As can be seen in Figure 4, NMOS transistors MN1 and MN2 form a current mirror. The size of MP1 is two times that of MP2. The gates of MP5 and MN3 are connected. Capacitance C1 and resistor R0 are frequency compensation devices. MP0 is a large pass transistor which enables heavier load capacity. MP4, MP5 and R2 stabilize the V_{CC} voltage level while C0 is used for filtering and storing energy. The voltage limiter of this circuit is composed of MP6, MP7, MN5, R1 and MN4.

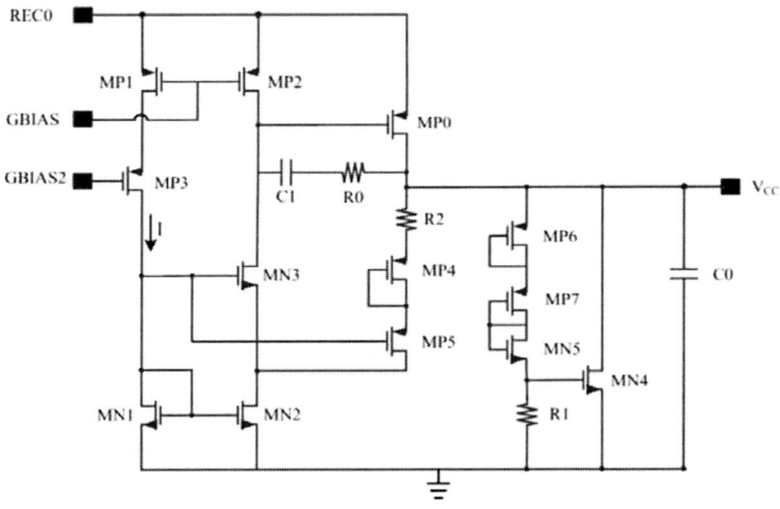

Figure 4. Circuit of V_{CC} regulator.

If calculating the output voltage through the way R2-MP4-MP5-MN1, the output voltage of the regulator in Figure 4 is given by [13]:

$$V_{CC} = V_{SG.MP4} + V_{SG.MP5} + I_{D.MP5} R_2 + V_{SG.MN1} \tag{6}$$

where MP4, MP5 and MN1 are in saturation region with the same W/L ratio, according to the V-I characteristic of MOSFET in saturation region, the following equation can be given as:

$$V_{SG,MP5} = V_{SG,MP4} = |V_{THP}| + \sqrt{\frac{2I_{D,MP5}}{\mu_P C_{ox}(W/L)_{MN1}}} \tag{7}$$

$$V_{SG,MN1} = V_{THN} + \sqrt{\frac{2I_{D,MN1}}{\mu_N C_{ox}(W/L)_{MN1}}} \tag{8}$$

V_{THN} and V_{THP} are the threshold voltages of NMOS and PMOS respectively.

As the size of MP1 is two times that of MP2, MN1 and MN2 are mirror transistors, and the current can be represented as:

$$2I_{D,MP5} = 2I_{D,MP2} = I_{D,MP1} = I_{D,MN1} \tag{9}$$

Equation (6) can be rewritten as:

$$V_{CC} = |V_{THN}| + 2|V_{THP}| + 2\sqrt{\frac{2I_{D.MP1}}{\mu_P C_{ox}(W/L)_{MN1}}} + \sqrt{\frac{2I_{D.MP1}}{\mu_N C_{ox}(W/L)_{MN1}}} + \frac{1}{2} I_{D.MP1} R_2 \tag{10}$$

In Equation (8) $I_{D,MP1}$ is determined by the bias circuit with the value of 2 μA. When ignoring the changes of temperature, the first and second term at the right side of Equation (10) are constant, whereas the last three terms are the only part that can be changed in this equation. Suppose that the last three terms of Equation (10) are equal. That is:

$$2\sqrt{\frac{2I_{D.MP1}}{\mu_P C_{ox}(W/L)_{MN1}}} = \sqrt{\frac{2I_{D.MP1}}{\mu_N C_{ox}(W/L)_{MN1}}} = \frac{1}{2} I_{D.MP1} R_2 = \frac{1}{3}(V_{CC} - |V_{THN}| - 2|V_{THP}|) \tag{11}$$

According to Equation (11), the value of $(W/L)_{MP1}$ and R_2 can be calculated under typical process conditions. The output voltage of the regulator changes along with the environment temperature and the rectified voltage of REC0. When the voltage of REC0 increases, $V_{SG,MP1}$ and $V_{SG,MP2}$ are clamped by the bias circuit. Since MP3 is a common-gate connection PMOS, the voltage change of REC0 mostly turns out to be the drain-source voltage of MP3, so the variation of drain voltage of MP1 and the current mirror mismatch of MP1 and MP2 caused by channel modulation effect are reduced. Similarly, the matching of MN1 and MN2 is improved by MN3. Thus, the immunity of the regulator output to voltage changes of REC0 are improved, and according to the analysis of the bias circuit mentioned later in Section 3.3, the current I of this regulator is not easily affected by the supply voltage of REC0, so fluctuations of V_{CC} caused by the supply voltage can be suppressed by a well-designed bias circuit.

Temperature Coefficient (TC) is an index which describes how temperature variations influence a specific value. It is mostly used to describe changes of resistance values. TC of this design is defined as Equation (12) V_1 and V_2 are voltages before and after the temperature change:

$$TC = \frac{V_2 - V_1}{V_1 \cdot \Delta T} \tag{12}$$

By observing the right side of Equation (10)V_{THN}, V_{THP}, $\mu_P C_{ox}$ and $\mu_N C_{ox}$ are process parameters and their contributions to V_{CC} are fixed when the W/L ratio is determined. TC of the fifth item $1/2I_{D,MP1}R_2$ can be modulated by selecting the proper type of resistor. TC of the front four items of Equation (10) around room temperature can be obtained by simulation. According to the obtained TC, a proper type of resistor for R2 can be selected. In this way, TC of V_{CC} is minimized around room temperature.

While types of resistors are limited in each process, the TC of resistors which can be chosen is also limited. To solve this problem, a concept of combination resistor is adopted in our design [19]. According to this concept, two resistors of different TC are chosen to compose a combination resistor. TC of V_{CC} can be minimized by choosing a proper TC of this combination resistor. As the TC of the combination resistor is between that of the chosen resistors, the combination resistor has many more TC options than a single resistor. In general, by setting a proper voltage proportion of the first three items in Equation (8) and selecting a combination resistor appropriately, a desirable V_{CC} voltage regulator circuit immune to temperature variations is designed in this work. Moreover, the circuit of the V_{DD} regulator has the same structure with that of V_{CC}, while loads of the output voltage are the main differences between the two, one can carry on the design and optimization of V_{DD} regulator with the same methods as V_{CC} mentioned above.

3.3. Bias Circuit

Shown in Figure 4, a bias circuit is designed to provide a stable current for the regulator and other circuits. Start-up part of this bias circuit is composed of PMOS transistors MP1 and MP2. In the process of starting up, if there are no transistors MP1 and MP2, the bias circuit will enter a condition that the voltage of GBIAS1 is high and the gate voltage of MN1 and MN2 is low when REC0 goes up. In this condition, there is no current in each branch of the bias circuit and the bias circuit isn't able to start to work. When starting up, the rectified voltage REC0 goes up and the voltage of GBIAS1 rises. At first, transistors MP1 and MP2 are both off and capacitor C1 is uncharged. Then MP2 turns on while REC0 goes high, thus making the voltage of GBIAS2 increase. Then transistors MN1 and MN2 turn on, reducing the drain voltage of MN2, thus making transistors MP5 and MP1 on. Capacitor C1 begins to charge through MP1. Hereafter, the bias circuit enters its normal working conditions when the upper plate of C1 is charged to REC0. Then MP2 turns off and the start-up circuit is out of operation afterwards.

As for the bias circuit itself, the current mirror in Figure 5 is composed of PMOS transistors MP4 and MP5. The W/L ratio of NMOS transistor MN2 is k times of MN1. MOS devices in Figure 5 are all in saturation region. As MP4 and MP5 are mirror transistors of each other, the current of each branch (represented as I_{bias}) is kept equal. There has the following equation [20]:

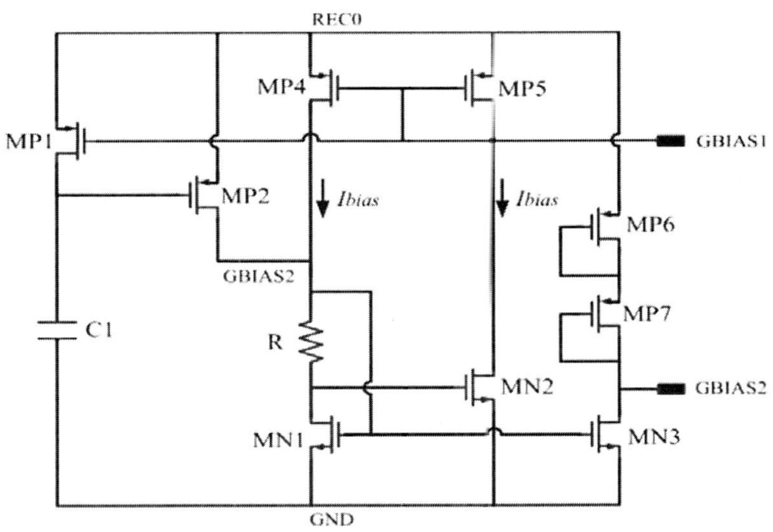

Figure 5. Bias circuit of RF powering circuit.

$$V_{GS,MN2} + I \cdot R = V_{SG,MP1} \tag{13}$$

Since it is known that MN3 and MN4 are in saturation region, according to the V-I characteristic of MOSFET in saturation region, I_{bias} can be calculated from Equation (13):

$$I_{bias} = \frac{2}{\mu_N C_{ox} (W / L)_{MN1} \cdot R_1^2} (1 - \frac{1}{\sqrt{k}})^2 \tag{14}$$

Shown in Equation (14), the advantage of this bias circuit is that it can eliminate the substrate bias effect. The W/L ratio of MN1, MN2 and the resistance value of R can be calculated according to the requirements of power dissipation, as well as the proper selection of over-drive voltage of MN1 and MN2.

As long channel devices are used in the bias circuit, the channel length modulation effect produced by the increasing of V_{RECO} can be well suppressed. Moreover, considering the temperature response of this circuit, Equation (15) can be obtained when taking the derivative of temperature in Equation (14):

$$\frac{\partial I}{\partial T} = A(\frac{1}{\mu_N} \cdot \frac{\partial \mu_N}{\partial T} + \frac{2}{R} \cdot \frac{\partial R}{\partial T}) \tag{15}$$

While $\partial\mu_N/\partial T$ is a fixed process coefficient, A is a fixed coefficient, and R represents the impedance of the resistor in Figure 5. By choosing proper value of $\partial R/\partial T$, the TC of bias current, expressed in Equation (15), can be minimized around room temperature. The combination resistor mentioned above is also adopted here.

3.4. Layout Optimization Methods

In this work, the RF powering circuit is designed for application in a passive RFID tag. As the tag is integrated with digital and analog parts in the same chip, the noise interference between the digital and analog parts becomes a serious problem as it operates under high frequency. As the RFID tag can work in intense electromagnetic field according to the related ISO/IEC 15693 and ISO/IEC 14443 standards, large spare energy in antennas must be bled by an over-voltage protection circuit, thus a large current is injected into the ground. At the same time, the high frequency of transient currents is conducted in the rectifier circuit.

These problems listed above can be latch-up triggers in both the analog and digital part, therefore leading to a latch-up in the chip which is a dominant failure in CMOS chip projects. Latch-up may cause a loss of data logic states or destructive failure of the whole system [21,22]. As a result, layout optimizations to isolate the noise between the analog and digital part and to reduce the possibility of failures caused by latch-up is quite essential in the layout design of this chip.

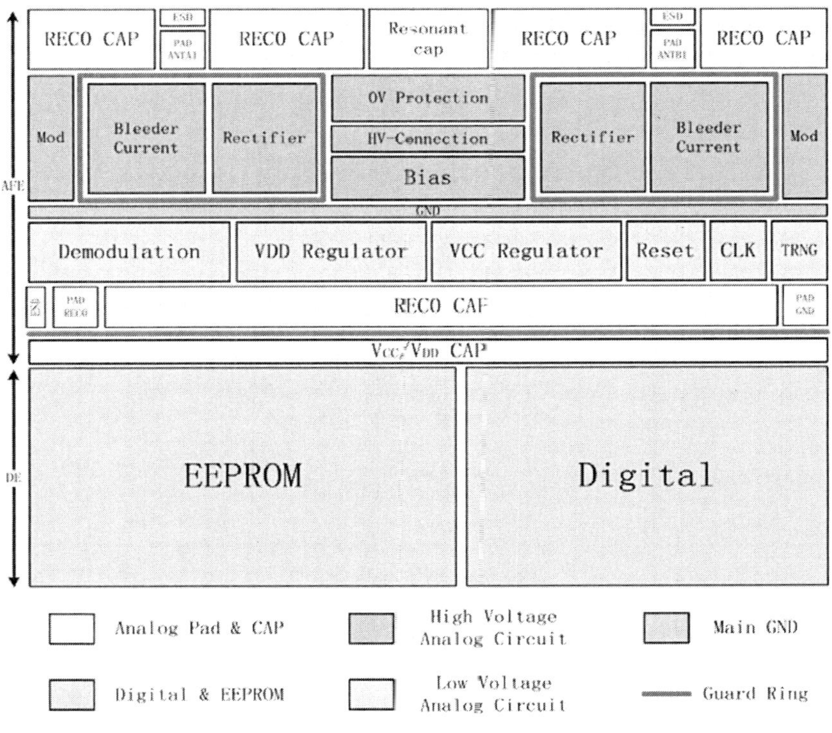

Figure 6. Layout floor plan of the whole chip.

The plan of the whole chip layout is shown in Figure 6. The AFE and DE part of the chip are separated with V_{RECO} and V_{CC}/V_{DD} capacitances so as to isolate noise between them. Since RECO is designed to be 4 V–12 V, which is much higher than the operation voltage of DE, an NP guard ring is placed between REC0 and V_{CC}/V_{DD} capacitances to split the high voltage AFE and low voltage DE areas. Moreover, in the AFE layout, high voltage blocks are apart from low voltage blocks with a main thick ground wire to avoid latch-up. To isolate substrate current in areas with large current flow, an NPN guard ring is added around the current bleeding and rectifier circuits for large currents injected into the substrate that increase the ground voltage which is more likely to lead to a latch-up.

For more considerations, two antenna pads ANTA1 and ANTB1 are placed as mirror images in this chip layout shown Figure 6, as well as wires, devices and circuits connected to the two antennas. Signals transferred through the two antennas are ensured to be symmetrical in this way. Buffer cells are added into the long wires connecting the AFE and DE circuits to suppress signal attenuations.

4. SIMULATION AND MEASUREMENT RESULTS

The RF powering circuit is designed and simulated in the HJTC 0.25 μm process together with the whole RFID tag. As mentioned in Section 3, PCE of the RF powering circuit designed in HJTC 0.25 μm process is simulated under different load current of V_{CC}/V_{DD}. Simulation results are shown in Table 2, PCE of the circuit is 39.17% when the load current is 520 μA while the tag is in its normal working conditions.

Table 2. Simulation results of PCE of RF powering circuit in HJTC 0.25 μm process.

V_{REC0}	I_{load}	$V_{ant,p\text{-}p}$	PCE
4.42 V	800 μA	7.03 V	53.14%
4.88 V	600 μA	7.36 V	43.68%
5.01 V	520 μA	7.43 V	39.17%
5.2 V	400 μA	7.53 V	31.94%

Temperature rejection ratio (TRR) and voltage rejection ratio (VRR) are defined as follows:

$$TRR = \frac{\Delta V_{CC}}{Temper} \tag{16}$$

$$VRR = \frac{\Delta V_{CC}}{V_{olt}} \tag{17}$$

where ΔV_{CC} is the variation of regulator output, Temper is the variation range of temperature; V_{olt} is the variation range of rectifier output voltage.

Shown in Figure 7 are the simulation results of regulator V_{CC} in the HJTC 0.25 μm process. The calculations of the simulation results are shown in Table 2. The value of PCE can be calculated with the equation PCE = P_{out}/P_{in}, while $P_{in} = \overline{V_{ant} \cdot I_{ant}}$ and $P_{out} = \overline{V_{CC} \cdot I_{load}}$ where V_{ant} is the voltage of the antenna, I_{ant} is the current of the antenna, V_{CC} is the output voltage of the regulator, I_{load} is the load current of the regulator. Considering the usual work environment of the tag, we pay more attention to the value of TRR while V_{RECO} is near 6 V, and the value of VRR while the temperature is near 27 °C. Under these two conditions the regulator has a quite small TRR and VRR with TRR = 0.04 mV/°C and VRR = 1.1 mV/V.

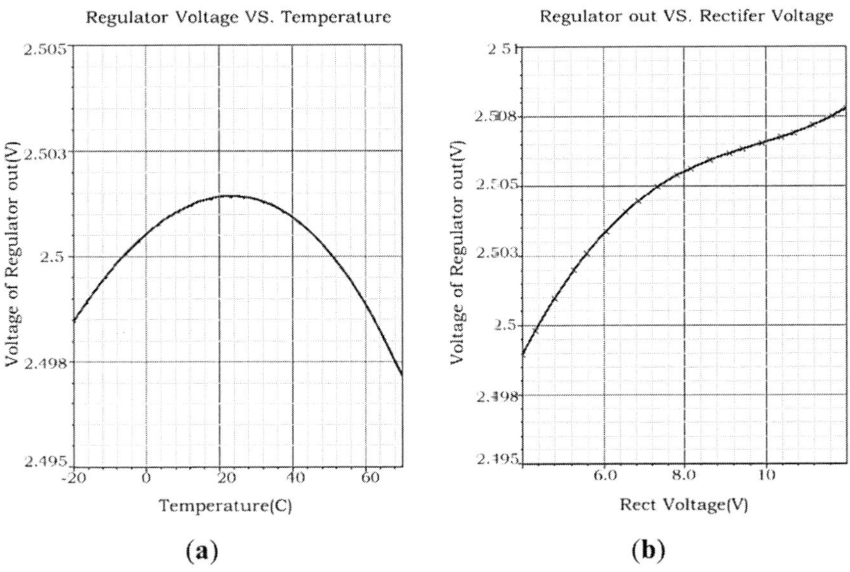

Figure 7. (**a**) V_{CC} vs. Temp Simulation results of regulator V_{CC}, V_{RECO} = 6V; (**b**) V_{CC} vs. V_{RECO} Simulation results of regulator V_{CC}, T = 27 °C.

RFID tags designed in this paper are also measured according to ISO.IEC 15693 standard and related ISO/IEC 10373-7 test standard with contactless test methods. The test electromagnetic field strength (represented as H) defined in the standard is 150 mA/m (H_{min} = 150 mA/m) to 5 A/m (H_{max} = 5 A/m), and test simulation signals are in 100% ASK or 10% ASK modulation mode.

The RF powering circuit has been applied to HF RFID tag chip projects. To make the testing more convenient, the analog front end (AFE) circuit for a HF RFID tag is fabricated, in which this RF powering circuit is integrated. Consisting of analog front end (AFE), digital baseband and EEPROM memory (DE), the RFID tag test platform based on PCB is shown in the Figure 8. The DE part is realized in the form of FPGA, the AFE part is fabricated in the HJTC 0.25 μm process. The RF powering circuit designed above is one part of the AFE in this chip. The partial enlarged view in Figure 8 shows that the distance from reader to the tag can be adjusted to test different electromagnetic field strengths. In Figure 8, the demodulated signal coming from the AFE chip and the return signal which is transformed from the demodulated signal just mentioned via the processing of the DE part are tested by the oscilloscope. In the oscilloscope screen shown in Figure 8, the upper curve is the test waveform of the demodulated signal, while the lower one is the test waveform of the return signal. Both of the test waveforms are normal, which means the communication between the tag and reader is successful. Thus we can affirm that the RF powering circuit can provide voltage source needed by the whole chip.

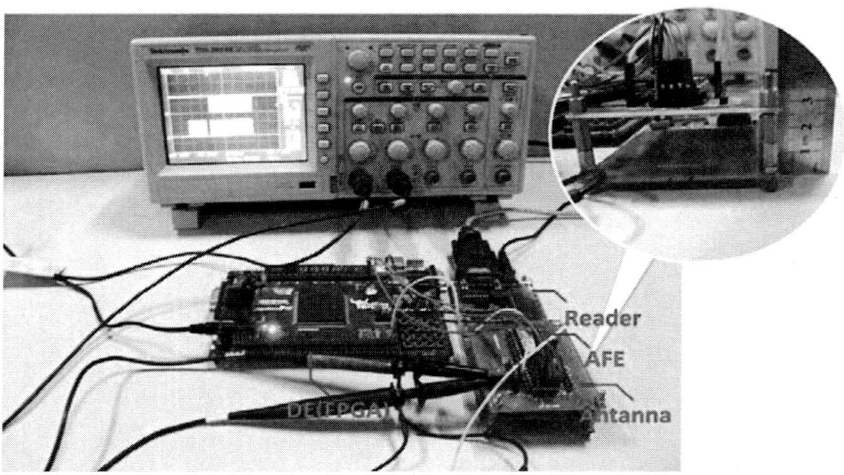

Figure 8. Functional measurement setup based on PCB and the demodulation signal from the tag.

The AFE was tested using a TDS 2024B oscilloscope. The test waveform of V_{CC} is shown in Figure 8. The strength of the electromagnetic field surrounding the RFID tag is adjusted by changing the distance between the chip and reader as shown in Figure 8. The V_{CC} test waveform when H = 0.3 A/m is shown in Figure 9a, and the one when H = 4 A/m is shown in Figure 9b. Other values of V_{CC} in different electromagnetic field are also tested, and they are shown in Figure 10. The test results of PCE of this RF powering circuit are shown in Figure 11.

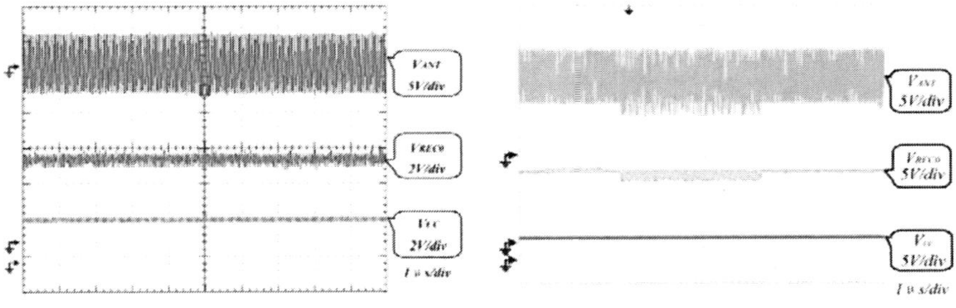

Figure 9. (**a**) V_{CC} test waveform when H = 0.3 A/m under 0.25 μm process, V_{CC} = 2.40 V (rms); (**b**) V_{CC} test waveform when H = 4 A/m under 0.25 μm process, V_{CC} = 2.42 V (rms).

According to Figure 9, the V_{CC} voltage level is 2.4 V within the normal deviation ±10% of 2.5 V. According to Figure 10, considering the usual work environment of the tag, the regulator of this RF powering circuit has a quite small VRR with VRR = 2.5 mV/V while V_{RECO} is near 6 V. According to Figure 11, considering the usual work environment of the tag, in which the input power of the RF powering circuit is near 0.32 mW, the PCE of the RF powering circuit is 38.54%. Table 3 shows the performance comparison between this RF powering circuit and previous RF powering circuits.

Table 3. Comparison with previous RF power circuits.

	This Work	[8]	[14]	[23]	[24]
Process	0.25 μm	0.35 μm	0.13 μm	0.35 μm	0.35 μm
VRR (mV/V)	2.5	39	12	4	0.8
TRR (mV/°C)	0.04	Not Support	0.2	0.01	2

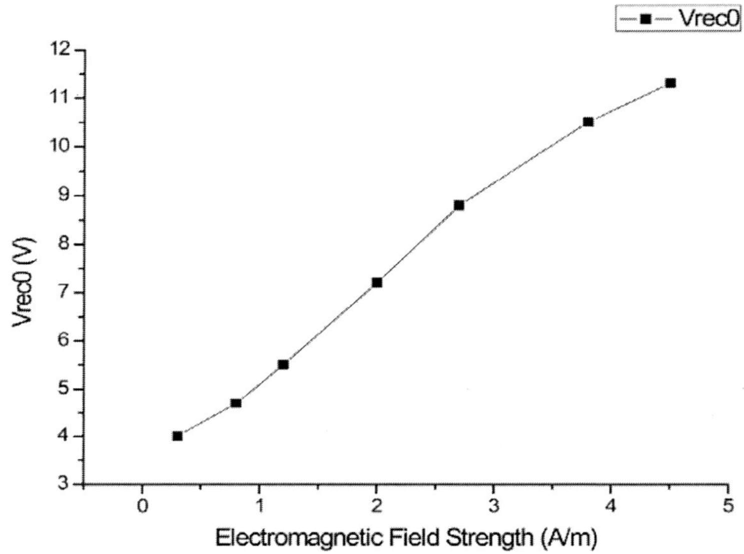

(a)

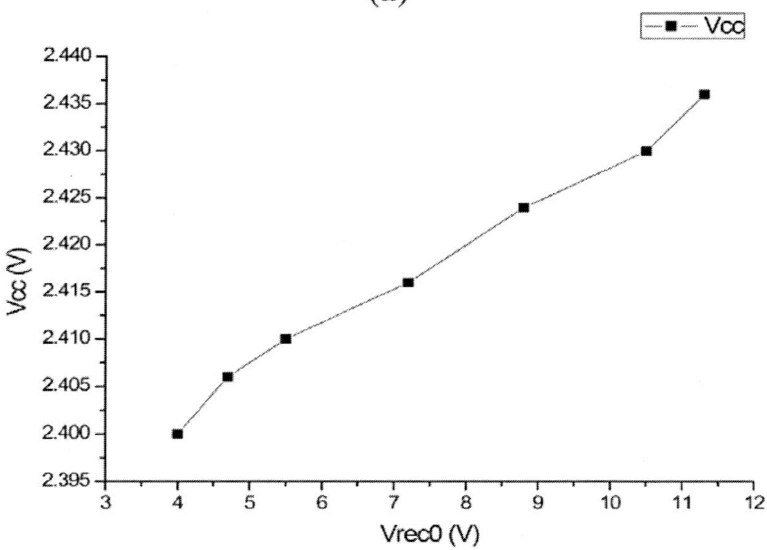

(b)

Figure 10. (**a**) V_{rec0} vs. Electromagnetic Field Strength; (**b**) V_{cc} vs. V_{rec0}.

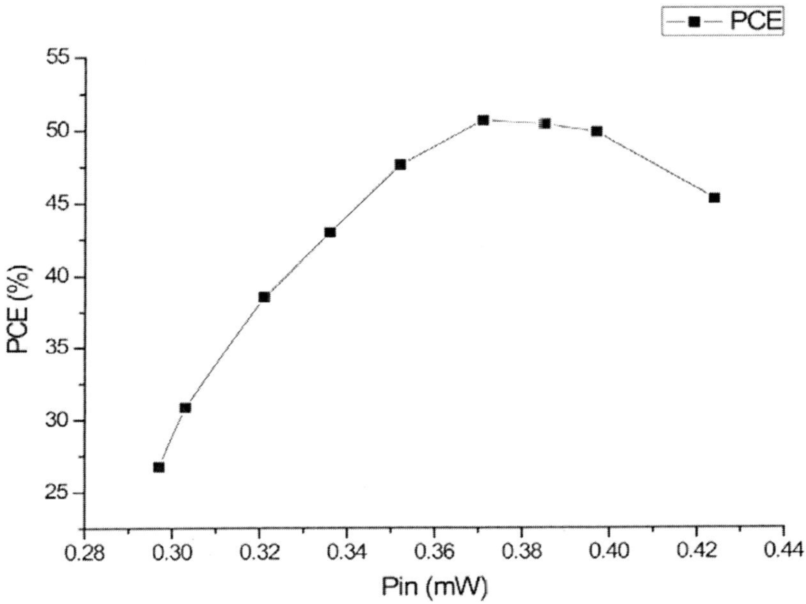

Figure 11. Test results of the PCE of the RF powering circuit in the HJTC 0.25 μm process.

Figure 12a shows that the demodulator of AFE has demodulated the downlink 10% ASK modulated signal sent from the reader successfully. The signal Demod10p is the demodulated signal achieved by the demodulator of AFE. Figure 12b shows an uplink signal of about 13.6% load modulated depth is achieved from the chip antenna to the reader which meets the communication criterion. The signal Datatran determined by the digital baseband controls the AFE modulator. Figure 13 shows that AFE has realized the information communication with reader successfully, which means this RF powering circuit work normally providing a stable voltage source for the whole chip.

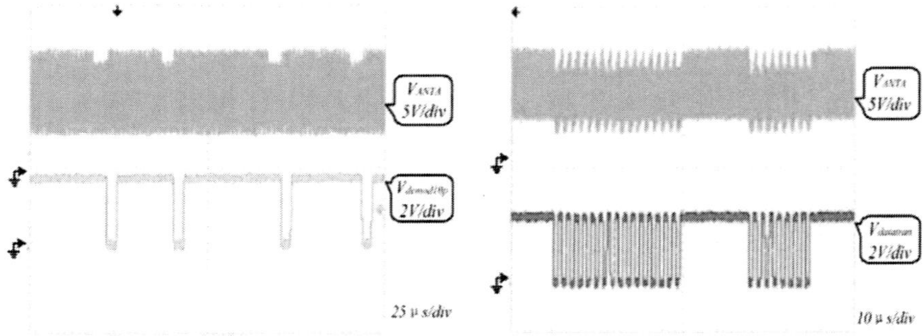

(a) (b)

Figure 12. (**a**) Waveform of demodulation; (**b**) Waveform of modulation.

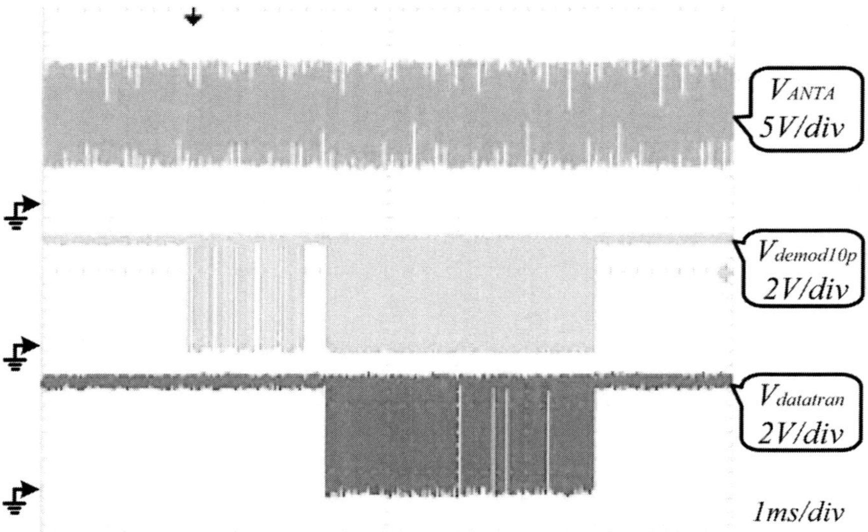

Figure 13. Measured waveform of the communication between reader and tag.

This RF powering circuit has been integrated in a RFID tag chip. The whole RFID tag packaged, chip die layout and chip photograph are shown in Figure 14. The size of the whole chip is 1.1 × 1.1 mm², whereas the RF powering circuit consumes 0.23 × 0.24 mm² of chip area. In Figure 14, there are also photographs of the packaged RFID tag and the tag used in a RFID smart card. The four pads for the package are two antenna pads and two voltage pads for testing.

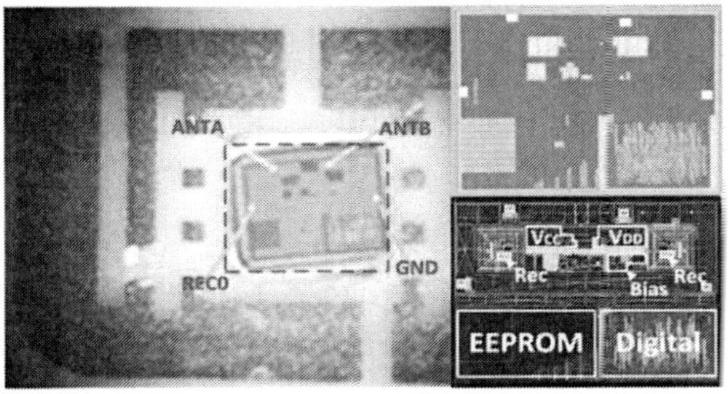

(a)

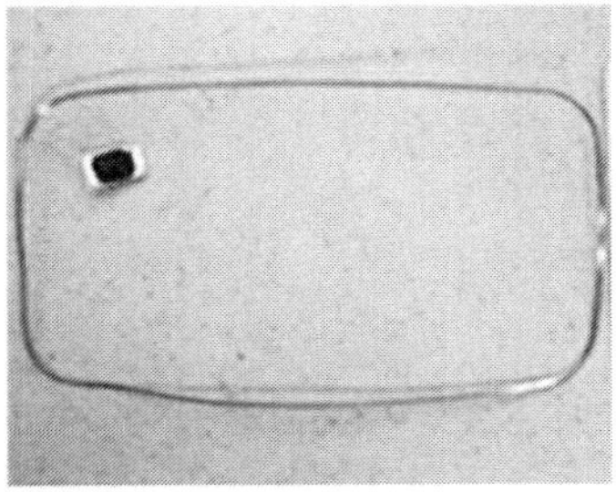

(b)

Figure 14. (**a**) Photograph of packaged RFID tag. Chip microphotograph and layout. (**b**) Chip tag used in a RFID smart card.

5. CONCLUSIONS

In this work, a RF Powering circuit used for passive RFID tags is designed and fabricated with the HJTC 0.25 μm CMOS process. The circuit consists of a rectifier, a bias circuit and a regulator. It provides two stable power sources for the analog and digital parts, respectively. The RF powering circuit achieves high PCE of 38.54% with the HJTC 0.25 μm process under normal working conditions and the power supply voltage is immune to temperature and voltage variations. Simulation results show that this circuit has a TRR = 0.04 mV/°C, and test results show that it has a VRR = 2.5 mV/V. This RF powering circuit is also fabricated with the HJTC 0.25 μm process in the ISO/IEC15693-Compatible and ISO/IEC 14443-Compatible RFID tag chip. Layout of this chip is optimized for latch-up prevention and noise isolation. Measurement results of the tag show that this circuit can provide a stable voltage source for a HF RFID tag chip for its proper operation.

ACKNOWLEDGMENTS

This work is supported by the National Natural Science Foundation of China (Grant No. 61306038 and No. 61376031), ChenGuang Project of science and technology for Wuhan youths (No. 2013072304010829) and the Fundamental Research Funds for the Central Universities (HUST: 2013TS050).

AUTHOR CONTRIBUTIONS

All authors have contributed to the presented work. Dongsheng Liu, Rencai Wang, Ke Yao, and Xuecheng Zou designed the whole circuit. Rencai Wang and Ke Yao conducted the implementation and wrote the manuscript. Liang Guo did the research on the regulator and the design of it. All authors have read and revised the manuscript.

CONFLICTS OF INTEREST

The authors declare no conflict of interest.

REFERENCES

1. Chiao-Tzu, H.; Shu-Jen, W.; Wei-Ling, W.; Yung-Sheng, W. Construction of an Online RFID Enabled Supply Chain System Reliability Monitoring Model. Proceedings of the 2012 International Symposium on Computer, Consumer and Control (IS3C), Taichung, Taiwan, 4–6 June 2012; pp. 626–629.
2. Sangdo, P.; Hongchul, L. Self-Recognition of Vehicle Position Using UHF Passive RFID Tags. IEEE Trans. Ind. Electron. **2013**, 60, 226–234.
3. Watfa, M.K.; Suleman, U.; Arafat, Z. RFID system implementation in Jebel Ali port. Proceedings of the 2013 IEEE Consumer Communications and Networking Conference (CCNC), Las Vegas, NV, USA, 11–14 January 2013; pp. 950–955.
4. Noorsal, E.; Sooksood, K.; Hongcheng, X.; Hornig, R.; Becker, J.; Ortmanns, M. A Neural Stimulator Frontend with High-Voltage Compliance and Programmable Pulse Shape for Epiretinal Implants. IEEE J. Solid-State Circuits **2012**, 47, 244–256.
5. Cazeca, M.J.; Mead, J.; Chen, J.; Nagarajan, R. Passive wireless displacement sensor based on RFID technology. Sens. Actuators A Phys. **2013**, 190, 197–202.
6. Rodriguez-Rodriguez, J.A.; Delgado-Restituto, M.; Masuch, J.; Rodriguez-Perez, A.; Alarcon, E.; Rodriguez-Vazquez, A. An Ultralow-Power Mixed-Signal Back End for Passive Sensor UHF RFID Transponders. IEEE Trans. Ind. Electron. **2012**, 59, 1310–1322.
7. Roberts, N.E.; Wentzloff, D.D. A 98 nW wake-up radio for wireless body area networks. Proceedings of the 2012 IEEE Radio

Frequency Integrated Circuits Symposium (RFIC), Montreal, QC, Canada, 17–19 June 2012; pp. 373–376.

8. Crepaldi, P.C.; Pimenta, T.C.; Moreno, R.L.; Rodriguez, E.C. A Low Power CMOS Voltage Regulator for a Wireless Blood Pressure Biosensor. IEEE Trans. Instrum. Meas. **2012**, 61, 729–739.
9. Ferreira, P.M.; Bergeret, E.; Pannier, P. 1.4 V and 300 nA UHF passive RFID voltage regulator. Proceedings of the 2012 IEEE 10th International New Circuits and Systems Conference (NEWCAS), Montreal, QC, Canada, 17–20 June 2012; pp. 429–432.
10. DiGiampaolo, E.; Martinelli, F. A Passive UHF-RFID System for the Localization of an Indoor Autonomous Vehicle. IEEE Trans. Ind. Electron. **2012**, 59, 3961–3970.
11. Chelho, C.; Young-Han, K.; Tae-Hun, K.; Kyusung, B.; Jongbae, K. Fully integrated ultra-low-power passive UHF RFID transponder IC. Proceedings of the 2011 IEEE International Symposium on Radio-Frequency Integration Technology (RFIT), Beijing, China, 30 November–2 December 2011; pp. 77–80.
12. Lijun, L.; Gannes, K.D.; Fricke, K.; Senjuti, S.; Sobot, R. Low-Power CMOS Voltage Regulator Architecture for Implantable RF Circuits. Proceedings of the 2012 Fourth International EURASIP Workshop on RFID Technology (EURASIP RFID), Torino, Italy, 27–28 September 2012; pp. 99–106.
13. Jong-Wook, L.; Duong, H.T.V.; Quoc-Hung, H.; Sang-Hoon, H. A Fully Integrated HF-Band Passive RFID Tag IC Using 0.18-μm CMOS Technology for Low-Cost Security Applications. IEEE Trans. Ind. Electron. **2011**, 58, 2531–2540.
14. Chi-En, L.; Yi-Jhan, H.; Jean-Fu, K. RFID Regulator Design Insensitive to Supply Voltage Ripple and Temperature Variation. IEEE Trans. Circuits Syst. II Express Briefs **2010**, 57, 255–259.
15. Liu, D.; Li, F.; Zou, X.; Liu, Y.; Hui, X.; Tao, X. New Analysis and Design of a RF Rectifier for RFID and Implantable Devices. Sensors **2011**, 11, 6494–6508.
16. Broutas, P.; Contopanagos, H.; Kyriakis-Bitzaros, E.D.; Tsoukalas, D.; Chatzandroulis, S. A low power RF harvester for a smart passive sensor tag with integrated antenna. Sens. Actuators A Phys. **2012**, 176, 34–45.
17. Scorcioni, S.; Larcher, L.; Bertacchini, A. Optimized CMOS RF-DC converters for remote wireless powering of RFID applications. Proceedings of the 2012 IEEE International Conference on RFID (RFID), Orlando, FL, USA, 3–5 April 2012; pp. 47–53.
18. Qiuyun, F.; Gay, N.; Bansleben, C.; Hildebrand, R. Optimization of the NMOS and PMOS gate cross-connected rectifier for RF power extraction in RFID applications. Proceedings of the Asia-Pacific

Microwave Conference (APMC 2008), Macau, China, 16–20 December 2008; pp. 1–4.

19. Gregoire, B.R.; Moon, U. Process-independent resistor temperature-coefficients using series/parallel and parallel/series composite resistors. Proceedings of the IEEE International Symposium on Circuits and Systems (ISCAS 2007), New Orleans, LA, USA, 27–30 May 2007; pp. 2826–2829.
20. Phillip, E.A.; Douglas, R.H. CMOS Analog Circuit Design, 2nd ed.; Publishing House of Electronics Industry: Beijing, China, 2011; pp. 109–124.
21. Momo, N.; Higashi, Y.; Oda, M.; Matsuzawa, K.; Kokubun, K.; Ohguro, T.; Momose, H.S.; Toyoshima, Y. 90 nm node RF CMOS technology with latch-up immunity on high-resistivity substrate. Proceedings of the European, Microwave Integrated Circuits Conference (EuMIC 2009), Rome, Italy, 28–29 September 2009; pp. 65–68.
22. Purwadi, A.; Shu-Ming, B.; Prabowo, B.A.; Jung-Ruey, T.; Gene, S. Shifting time waveform induced CMOS latch up in bootstrapping technique applications. Proceedings of the 2012 19th IEEE International Symposium on the Physical and Failure Analysis of Integrated Circuits (IPFA), Singapore, Singapore, 2–6 July 2012; pp. 1–4.
23. Minchin, L.; Chichun, H.; Zongwei, L. Implementation of Low Dropout Regulator with Low Bandgap Reference Voltage Circuit for RFID Tag Applications. Proceedings of the Cross Strait Quad-Regional Radio Science and Wireless Technology Conference (CSQRWC), New Taipei City, Taiwan, 23–27 July 2012; pp. 40–43.
24. De-Vita, G.; Iannaccone, G. Ultra-low-power series voltage regulator for passive RFID transponders with subthreshold logic. Electron. Lett. **2006**, 42, 1350–1351.

11 CHAPTER

DESIGN OF A COVERT RFID TAG NETWORK FOR TARGET DISCOVERY AND TARGET INFORMATION ROUTING

Qihe Pan and Ram M. Narayanan *

Department of Electrical Engineering, The Pennsylvania State University, University Park, PA 16802, USA

ABSTRACT

Radio frequency identification (RFID) tags are small electronic devices working in the radio frequency range. They use wireless radio communications to automatically identify objects or people without the need for line-of-sight or contact, and are widely used in inventory tracking, object location, environmental monitoring. This paper presents a design of a covert RFID tag network for target discovery and target information routing. In the design, a static or very slowly moving target in the field of RFID tags transmits a distinct pseudo-noise signal, and the RFID tags in the network collect the target information and route it to the command center. A map of each RFID tag's location is saved at command center, which can determine where a RFID tag is located based on each RFID tag's ID. We propose the target information collection method with target association and clustering, and we also propose the information routing algorithm within the RFID tag network. The design and operation of the proposed algorithms are illustrated through examples. Simulation results demonstrate the effectiveness of the design.

Keywords:RFID tag network; pseudo-noise signal; cluster; routing

1. INTRODUCTION

RFID tags are small electronic devices working in the radio frequency (RF) range. They use wireless radio communications to automatically identify objects or people without the need for line-of-sight or contact, and have the advantage that they can be read through a variety of visually and environmentally challenging conditions. Their properties such as low cost, small size, and wireless functioning make them widely used in inventory tracking, object location, environmental monitoring, etc. Based on their energy source, RFID tags are categorized into three types: passive, semi-passive, and active. Passive RFID tags use energy from the incoming signal to power themselves, while semi-passive and active RFID tags use an internal power source, usually a small battery. Thus, active RFID tags can perform advanced functions and also work over longer ranges. RFID is an exciting area for research due to its relative novelty and exploding growth. Current research on RFID focuses on RF tags, readers, communication infrastructure, as well as some policy and security issues [1]. This paper explores the applications of active RFID tags in target identification in the RFID tag network using a pseudo-noise signal, where a static or slowly moving target out of the range of the command center transmits a distinct pseudo-noise signal within the field of the spatially distributed RFID tags. These RFID tags in the network collect the target's information and route it to the command center. The noise signal from the target is known only to the RFID tags in the network so they can easily detect it. However, this signal is not detected by undesired parties since the transmitted signal has unpredictable random-like behavior and does not possess repeatable features for signal identification purposes [2]. Noisy tags, which are regular RFID tags that generate noise, can be used to help establish a secure channel between the reader and the queried tag. A noisy tag protocol is proposed in [3], wherein a noisy tag in the reader's field sends out a noise signal generated from a pseudo-random function, the secret shared with the reader. The reader can reconstruct and subtract the noise signals from the noisy tag and recover the message from the queried tag, while an eavesdropper is unable recover the queried tag's message. An eavesdropping-resistant and privacy-friendly RFID system is developed in

[4], in which the chip modulates its reply onto a noisy carrier provided by the reader to protect the back-channel against eavesdropping. This method does not require additional protective devices.

Cluster approaches have been used a lot in parametric frameworks for detection and estimation. Sensors are partitioned into subgroups for distributed learning in the wireless networks [5]. In addition, the cluster approach is also used for topology control [6]. In this paper, we employ RFID tag clusters within the RFID tag network to collect the target's information. There are two steps in this process: (1) target association; and (2) cluster formation and cluster head selection. If an RFID tag detects the target, then it stores the target's ID and gets associated with the target. Clusters are formed by RFID tags associated with the same target. One of these RFID tags, selected as the head of the RFID tag cluster, routes the target's information out to the command center. In our design, the RFID tag with the maximum number of links to the outside of the cluster is selected as cluster head, which is robust to channel failures, considering that the RFID tags in the network are battery driven and may run out of life. When some of the communication links between the cluster head and those RFID tags out of the cluster are broken, the cluster head RFID tag still can use alternate communication links between it and RFID tags outside of the cluster to route the target's information out.

There are many approaches for information routing in the wireless sensor networks from different aspects of view. In [7], an information-directed routing method is proposed for localization and tracking problems, in which routing is formulated as a joint optimization of data transport and information aggregation, and information accumulated is maximized along the routing path. In [8], selection of the set of cluster heads is defined as the weighted connected dominating set problem, and centralized approximation algorithms are developed to select them. A maximum energy welfare algorithm is designed in [9] by applying the social welfare functions to the routing in wireless sensor networks. Each sensor makes routing decisions to maximize the energy welfare of its local society, which leads to globally efficient energy-balancing due to overlapping of the local societies. RFID tags can also be used to route information in the networks. In [10], the active relay tags retransmit their received signals during the communication between the interrogator and active tags, and the proposed RFID multi-hop relay system can achieve larger coverage. In our approach, the routing path in the RFID tag network from cluster head RFID tag to the command center is selected according to the channel condition, which is a joint optimization of favorable channel conditions and short path length. Each RFID tag intelligently selects its successor and routes the target's information to it. There are two stages when each node selects its successor on the routing path based on two criteria: (1) channel quality sensing; and (2) target's information routing. During channel quality sensing, the channel condition is estimated and quantized to form the link weight, while in the information routing stage, the RFID tag determines its successor based on the channel information obtained and sends the target's information to it.

In this paper, we present an algorithm design in the physical layer on target information collection and routing within the RFID tag network in outdoor scenarios, specify the signal format, signal modulation, and signal detection method. Using a noise signal as the information carrier and a noisy key at the front of the RFID tag's signal indicating the purpose of the message, guarantees that the communication within the RFID tag network is covert, owing to the low probability of interception and low probability of detection of the noise waveform. During the RFID tag cluster head selection process, the RFID tag with the maximum number of links to the outside of the cluster is selected as the tag cluster head, and it routes the target information out to the command center. The RFID tag cluster head selected in this manner is robust to channel failures. When some of the communication links between it and the RFID tags out of the cluster turn down, which may occur due to the battery failure in those RFID tags, it still can use the other communication links between it and RFID tags outside of the cluster to route the target's information out. The routing path from RFID tag cluster head to the command center in the RFID tag network we propose is based on the joint optimization of channel quality and path length.

The rest of the paper is organized as follows. Section 2 gives the design of the RFID tag network and procedures for target's information collection in the RFID tag network. Section 3 presents the algorithm for target's information routing within the RFID tag network, and it is illustrated through examples. In Section 4, we discuss implementation issues for hardware realization. Section 5 draws the conclusions of this article and presents possible future extensions.

2. TARGET INFORMATION COLLECTION

The application scenario is depicted in Figure 1. A static or slowly moving target is in the field of RFID tags, out of the range of the command center. The cooperative target transmits a distinct RF pseudo-noise signal. The goal of the RFID tag network design is to collect the target's information and route it to the command center with assistance of the deployed RFID tags. The RFID tags here do not know their own locations. A map of their

locations is saved at the command center, so the command center can determine where a RFID tag is located based on the RFID tag's ID.

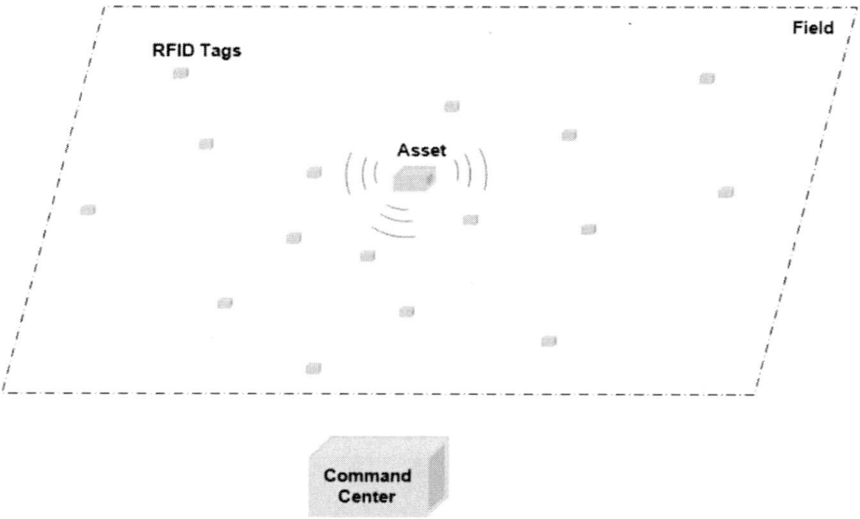

Figure 1. RFID tag application scenario.

In this stage, there are two steps for collecting the target's information: (1) target association; and (2) cluster formation and cluster head selection. In target association, some of the RFID tags detect the target by sensing the environment and record the target's information, and thus these RFID tags are associated with that target. RFID tags associated with the same target form tag clusters. Within a cluster, RFID tags share the same information associated with the target, so when some of the RFID tags turn down due to battery failure, etc., other RFID tags still have the target's information. One RFID tag, namely the head of the RFID tag cluster needs to route the target's information out. The cluster head RFID tag is chosen during cluster head selection.

2.1. Target Association

The target under monitoring transmits its distinct signal in noise form in the RFID tag field. The RFID tags designed here have templates of the signals from possible targets of interest. They listen to the environment and detect whether there is any target that is on the monitor list of potential targets. Each RFID tag recognizes the target by comparing its received signal with its template signals in its memory. Once an RFID tag detects a target, it records the target's information, such as the target's ID. Since signal transmitted by the target is of random noise, RFID tags use the cross-correlation process to determine whether the target exists in the field or not. In the real world, the environment is more complex with various interferences such as clutter, Doppler shifts, etc., which are not fully discussed here since they are not the main focus of the paper.

Suppose the target transmits pseudo-random noise signal burst s(t) over time T_0, and RFID tag has the library of signals from possible targets on the list $\{s_i(t)\}$,i= 1, 2,...,M, where M is the number of targets in the monitoring list. That is, in the RFID tag's library, signal $s_i(t)$ is a template of the signal transmitted by the ith target. The detection output at the RFID tag is:

$$corr(\tau) = \int_0^T s(t)s_i(t-\tau)dt, \quad for\ j = 1,2\ldots,M \tag{1}$$

If there is a peak at some time index of the correlation output, it means that the target's signal does exist, and therefore the target's ID is determined.

The RFID tag records the ID of the target that it is associated with to its memory variable $flag_{tag\_id}$, and modulates it to the tag's signal. By default, if an RFID tag is not associated with any target, its $flag_{tag\_id}$ is 0. The RFID tag's signal has the general format shown below (Figure 2), where each section is denoted by the bits under it. For each section, the all-0 bit message means that the RFID tag's signal contains no specific information of that section.

Key	Target ID	Tag ID	Tag ID (o.c.)	Counter
2 bit	2 bit	6 bit	6 bit	3 bit

Figure 2. RFID tag's signal general format

Each RFID tag's base signal comes from filtering a band-limited pseudo-noise signal $s_{tag}(t)$ to a specified and unique frequency sub-band. Different RFID tags have non-overlapping frequency bands, and they all have knowledge of $s_{tag}(t)$ in advance. For example, RFID tag k's base signal $s_{tag\_kb}(t)$ is $s_{tag}(t)$ filtered to its kth sub-band, and RFID tag k's signal $s_{tag\_k}(t)$ modulated with message is described as:

$$s_{tag\_k}(t) = \sum_{n=0}^{L-1} a_n s_{tag\_kb}(t - nT) \tag{2}$$

where T is the time duration of $s_{tag\_kb}(t)$, n is the index of the bit, a_n is the valued of the nth bit, and L is the number of total bits of RFID tag's signal.

After association with a target, the RFID tag's signal has the format shown in Figure 3, where Target ID denotes the target's ID that the RFID tag is associated with.

×	Target ID	Tag ID	×	×

Figure 3. RFID tag's signal format after association with a target.

If a target is in the field, only a subset of the RFID tags can collect its information. This is due to the fact that the distance between the target and the RFID tags may be larger than the detection range of some of the tags, or that the channel condition is very bad due to excessive noise making the error probability from that link above the tolerance level.

The target association process of an RFID tag is illustrated with simulations in Figure 4. In the simulation, the target's signal is assumed to be over the 1–2 GHz frequency band. We assume also that there are 50 RFID tags in the field, the pseudo-noise signal $s_{tag}(t)$ is over 1–2 GHz, and the RFID tag (ID 10)'s signal is over the 1–1.0187 GHz sub-band. RFID tag (ID 10) detects and gets associated with the target (ID 01) in a channel with a signal-to-noise ratio (SNR) of −3 dB. The negative SNR shows that the target association process is performed covertly since the signal power is less than that of the channel noise.

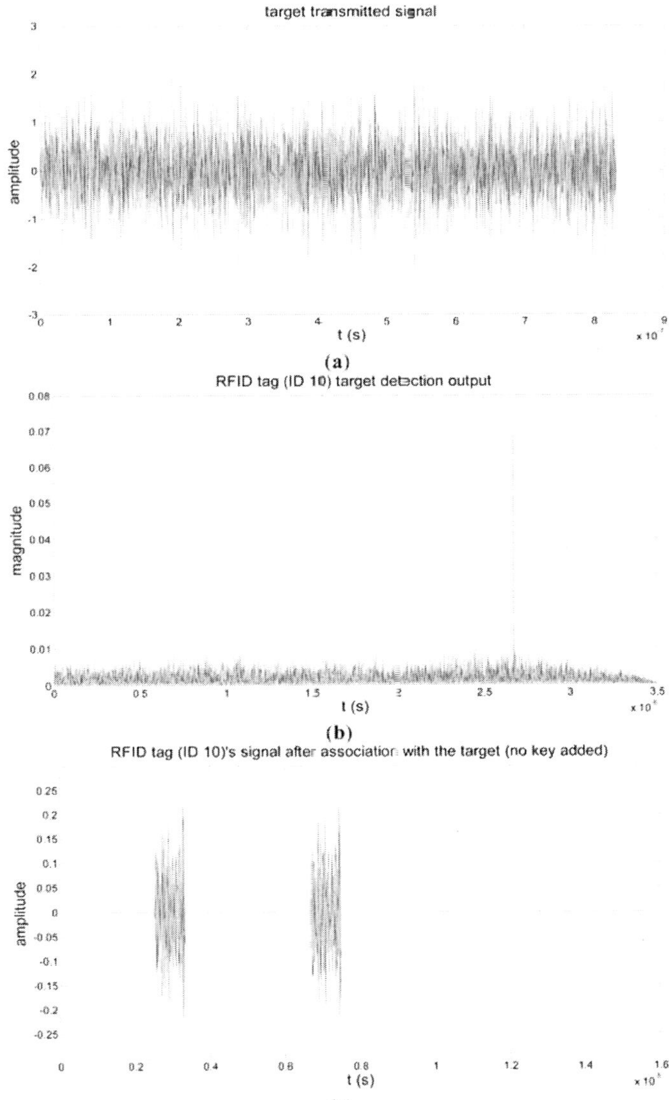

Figure 4. Target association simulation illustration. (**a**) noise signal transmitted by the target; (**b**) correlation output of RFID tag (ID 10) indicating target detection; (**c**) RFID tag (ID 10)'s signal after association with a target with no key.

2.2. Cluster Formation and Cluster Head Selection

RFID tags associated with the same target are formed as a cluster. In each cluster, cluster head RFID tag is selected through inter-communication among the cluster member RFID tags, and is responsible for routing the target's information the cluster's associated with to the outside of the cluster. Using cluster head RFID tag to route the target's information out is to reduce the information redundancy and signal interferences from multiple RFID tags.

RFID tags here are power driven devices, so the links between them may fail occasionally. To ensure connectivity, the RFID tag with the maximum number of links to the outside of the cluster is selected as cluster head, which is responsible for routing the target's information it carries to the outside of the cluster. The cluster head RFID tag selected accordingly is robust to channel failures. When some of the communication links between it and those RFID tags out of the cluster turn down, the cluster head RFID tag still can use the other communication links between it and RFID tags outside of the cluster to route the target's information out. When a tie occurs, i.e., when two or more concurrent RFID tags have the same number of links to the outside of the cluster, the one with the highest energy level is selected as the cluster head.

We model the RFID tag network as a graph G(V,E), where V is the set of nodes in the graph G, and E is the set of edges. The cluster of RFID tags is modeled as a subgraph C of G. Each RFID tag is represented by a node in the graph and the communication channel between RFID tags is represented by an edge. Then, the cluster head RFID tag is the start node on the routing path of the target's information with which all RFID tags in the cluster are associated.

In our system, the RFID tag is designed to operate in two modes. In Mode I, the default mode, the RFID tag works at normal energy level. In Mode II, the RFID tag works at higher energy and has longer communication distance. Most of the time, the RFID tags operate in Mode I. In the case an RFID tag needs a larger range, for example, when it tries to find its neighbors but cannot find any in the default mode, the RFID tag will go to Mode II. When the task is finished, the RFID tag will return to the low-energy Mode I.

For the design of the RFID tag's operating Mode II, we assume that each cluster of RFID tags is a connected component in the RFID tag network. That is, for each pair of nodes $u, v \in V(C)$, there is a u,v -path in C. Thus, RFID tags within the same cluster are able to get messages of the rest in the cluster tags. From these messages, the RFID tags recognize other member RFID tags in their cluster and the cluster head RFID tag is determined.

After sensing and association with the target, the RFID tag starts to discover and count its links with RFID tags not associated with the target. The RFID tag associated with the target sends out its outside-link sensing signal of the format shown in Figure 5.

Key (1)	Target ID	Tag ID	×	×

Figure 5. RFID tag's inquiry signal format for counting outside links.

The Key at the front of the RFID tag's message is globally defined, known by all the RFID tags in the field, to indicate the purpose of the RFID tag's message. Here, Key (1) indicates that the RFID tag which sends out the message is sensing and counting its links with RFID tags outside the cluster. For signal covertness, the Key is designed to be a noise waveform. RFID tags in the network have templates of the Keys, and they can recognize the corresponding Keys by cross-correlating the incoming signal with their stored templates of Keys.

If an RFID tag hears the link inquiry from one RFID tag within the cluster, it obtains the Key in the message and determines the type of the Key. If the Key is Type 1, it decodes the message to get the target's ID. In addition, it checks whether its own signal has that particular target's ID stored. If it is not associated with the target, it sends back the signal modulated with its ID; else, it does not respond. This guarantees that the RFID tag in the cluster only counts its links with RFID tags outside the cluster. The RFID tag outside the cluster responds to the cluster member RFID tag with the signal format shown in Figure 6 upon the link counting inquiry, where Tag ID (o.c.) denotes the RFID tag's ID outside the cluster.

Key (1)	Target ID	Tag ID	Tag ID (o.c.)	×

Figure 6. RFID tag's response signal format for counting outside links.

The RFID tag stores the number of its links with RFID tags outside the cluster in a counter, which is set to zero (0) by default. RFID tag sensing links with those outside the cluster obtains and determines whether the Key in the message is Type 1 upon its received signal, if so, it decodes the message. It checks whether the first Tag ID in the message is the same as its own to determine whether the message is a response to its link counting inquiry. Then it increases the number of links in its counter by 1 if there is a new RFID tag ID in the message. After searching for links to the outside of the cluster, the RFID tag updates its signal following the format in Figure 7, where Counter saves the number of links the RFID tag has to the outside of the cluster, and Key is set to initial value which is blank and has no meaning about the function of the message.

Key	Target ID	Tag ID	×	Counter

Figure 7. RFID tag's signal format after searching for links to the outside of the cluster.

After time Δt_1 from the time it sends out the link counting inquiry signal, the RFID tag stops receiving the responses to its link counting inquiry, and finishes counting the number of links it has with RFID tags outside the cluster. An approximate Δt_1 is given as:

$$\Delta t_1 \approx \frac{2R}{c} \tag{3}$$

where R is the range of RFID tag, and c is the speed of light in the air.

If the RFID tag cannot find any neighbor outside the cluster at this time, it changes to operation Mode II, and starts searching the links again. In Mode II, the RFID tag functions with more energy than in the default mode. With the design of Mode II, we assume that at least one RFID tag in the cluster has positive Counter. The RFID tag returns to the default operation mode after completing searching its links to the outside of the cluster.

The RFID tags in the cluster that complete the whole searching for links in Mode I wait for time Δt_1 from the time they finish searching for links. Thus, all the RFID tags in the cluster spend the same time $2\Delta t_1$ on the process to search for links to the outside of the cluster. The link counting process for RFID tag associated with a target is depicted in Figure 8.

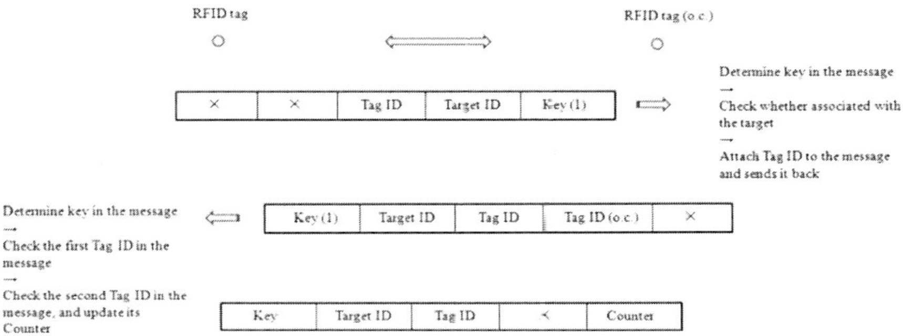

Figure 8. Link counting process of RFID tag in the cluster.

Some of the RFID tags in the cluster may be closer to the target than others, so they send out the link count inquiry signals earlier, and complete the link searching and counting process described earlier. We set a time variable Δt_W[tag_id] for each RFID tag to wait after it completes the link searching process, before starting inter-cluster communication. Thus, when the RFID tags in the cluster start inter-cluster communication, they have all finished counting the links and each of their Counters stores the final values.

An approximate value for Δt_W[tag_id] is as follows:

$$\Delta t_w[tag\_id] \approx \frac{K}{t_T[tag\_id]} \tag{4}$$

where K is a constant, and t_T[tag_id] is the target discover time at that RFID tag. Thus, the RFID tag closer to the target wait for longer time after complete searching and counting its links to the outside of the cluster.

Then the RFID tags in the cluster starts inter-cluster communication to recognize the members in the cluster and select the cluster head RFID tag. Each RFID tag broadcasts its signal in the format shown in Figure 9, where Key 2 indicates that the message is communicated among RFID tags in the cluster to select the cluster head.

Key (2)	Target ID	Tag ID	×	Counter

Figure 9. RFID tag's format for inter-cluster communication.

Upon receiving the signal, the RFID tag in the cluster obtains the Key in the message and determines whether it is Key 2. If it is Key 2, the RFID tags with their $flag_{tag\_id}$ registered will involve in the inter-cluster communication, and those with $flag_{tag\_id}$ of 0 will not. In the case there is only one target in the field, this also indicates that the RFID tag is within the same cluster. In the complex case of multiple targets in the field, the RFID tag needs to further check the Target ID in the message to determine if it is the same as its own or not. If so, the message is from an RFID tag in the same cluster. After determining that the message is from an RFID tag in the same cluster, the RFID tag continues to decode the message and checks whether the Tag ID in the message is the same as its own. If the message is from another RFID tag for the purpose of cluster head RFID tag selection, it forwards the message and compares the Counter in the message with its own. If the Counter in the message is larger than its own, the RFID tag sets its own Counter to −1, which indicates that its number of links to the outside of the cluster has been compared and not the largest.

After sufficient time Δt_2, each RFID tag in the cluster completes deciding whether it has the most number of links to the outside of the cluster. The RFID tag whose Counter is positive will become the cluster head, and it then starts to route the target's information out. Then all the RFID tags' Counters will be initialized to zero.

An approximation for Δt_2 is given as follows:

$$\Delta t_2 = \frac{2\pi R - R}{c} \tag{5}$$

which is a little larger than the worst time of cluster head selection. This approximation for Δt_2 in Equation (5) is based on the case shown in Figure 10.

As stated before, the RFID tag network is modeled as a graph, where each node represents an RFID tag and each link represents the communication link between RFID tags. The cluster of RFID tags is then a subgraph, and it is a connected component in our assumption with the design of RFID tag's operation Mode II. In the case shown in Figure 10, the two black nodes are in the same cluster, but the distance between them exceeds their range, and they cannot communicate with each other directly. Since the cluster is a connected component, there exists a path in the cluster connecting the two nodes. Through message forwarding by other nodes in the cluster, the two black nodes can communicate indirectly, for example, following the route in dashed line in Figure 10. The route length is on the order of $2\pi R - R$, and this costs time on the order of $2\pi R - Rc$. Thus, the cluster head is the RFID tag with the maximum number of links to the outside of the cluster.

In a complex case, several RFID tags in the cluster have the same number of links to the outside of the cluster. Since each RFID tag is ignorant of its location, it does not know whether it is nearest to the command center or not. Thus, the RFID tags in the cluster are unable to select the one nearest to the command center among them as the cluster head. Instead, they may further communicate to select the one with most energy as the cluster head. In this paper, we restrict the situation to the simple case that there is no tie.

After a short time, when the target's information is routed to RFID tags that have no links with the RFID tags in the cluster, RFID tags in the cluster set their $flag_{tag\_id}$ to 0, return to the beginning state and starts a new cycle. They perform target association, cluster formation, and cluster head selection again.

Additionally, if we upgrade the RFID tag design, as shown in Figure 11, such that the cluster head RFID tag is capable of saving the IDs of other RFID tags in the cluster during the inter-cluster communication, and it incorporates that information to the message to be routed outside the cluster, the target's location can also be determined at the command center. As stated before, the command center has a map of all the RFID tags. Thus, if the IDs of at least three RFID tags associated with the target are known, the locations of these three RFID tags are known at the command center, and thereby the location of the target can be determined.

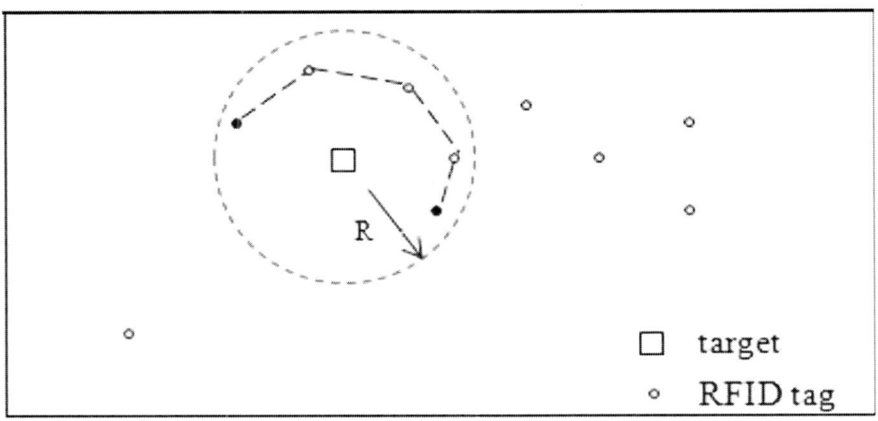

Figure 10. Case for maximum Δt_2.

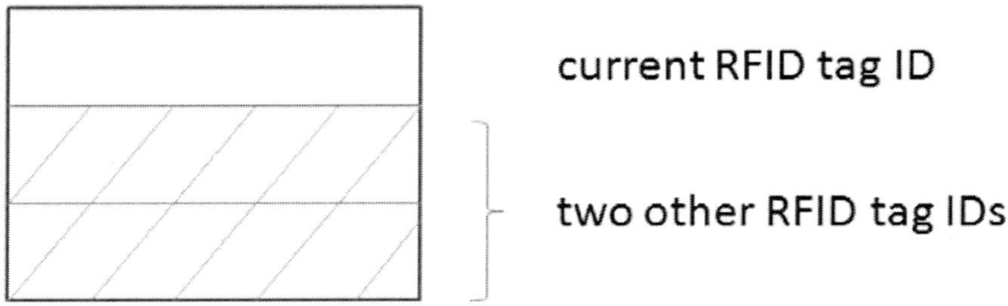

Figure 11. Signal format modification for target location determination.

3. INFORMATION ROUTING

The signal sent by an RFID tag is not of very high power. With the assumption that the communication cost is proportional to the communication distance, the goal of information routing in the RFID tag network is to select a channel which is robust and of short path length to route the target's information gathered by the RFID tags to the command center. When each node selects its successor on the routing path, there are two stages during the process: channel quality sensing and target's information routing. In channel quality sensing, the link weight is estimated based on the corresponding channel condition. In the information routing stage, the node determines its successor and sends the target's information to it.

The RFID tag network is modeled as a two dimensional graph G = (V,E), where V = $\{v_1,v_2,\ldots, v_n\}$ is set of the nodes, representing the RFID tags, and E is the set of bidirectional links, representing the communication links between RFID tags. Each link, shown in Figure 12, is assigned a positive weight which indicates the robustness of its corresponding communication channel. If the quality of the communication channel of is good, its weight is small; if the channel is bad, for example, very low SNR, excessive fading, object blocked channel, etc., its weight is very large; if there is no link between the two nodes, the weight is ∞. The weight for link (v_i, v_j) is expressed as w_{ij}.

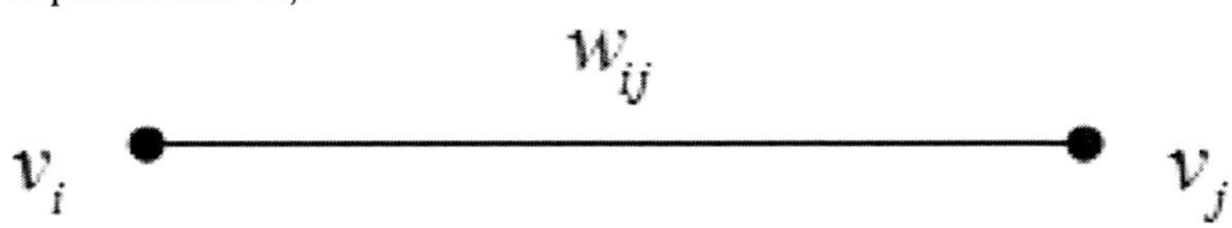

Figure 12. Model of a link.

3.1. Channel Quality Sensing

In this step, the RFID tag senses the channels. All its neighbor RFID tags calculate the weights of links connected to them based on their received signals, and respond to the RFID tag with updated messages. If a tag does not find any neighbor, it transmits in power Mode II. The RFID tag sends out the signal format shown in Figure 13 for channel sensing, where Key (3) indicates that the message is for channel sensing.

Counter (/ t(Vi))	×	Tag ID (i)	Target ID	Key (3)

Figure 13. RFID tag's signal format for channel sensing.

The channel quality sensing process between two RFID tag nodes is depicted in Figure 14. Since RFID tags in the cluster associated with the target have the target ID stored in their memory variables $flag_{tag\_id}$, when they receive the channel quality sensing message indicated by Key (3), they will not respond and thus will not be involved in the target's information routing. As for RFID tags outside the cluster, their memory variables $flag_{tag\_id}$ do not have the target's ID, and they will participate in routing the target's information.

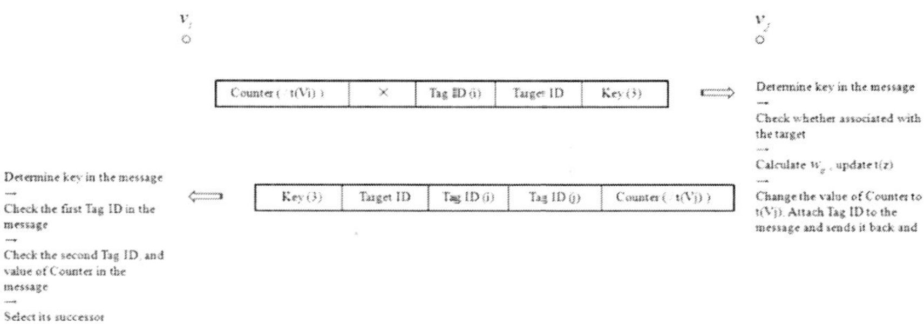

Figure 14. Channel quality sensing process between two RFID tag nodes.

In Figure 14, node v_j decodes the signal from node v_i, estimates the channel condition for link (v_i,v_j), quantizes it, and grades it to the link weight w_{ij}, generated by a channel quality quantization function. The value of Counter in the message is $t(v_x)$, where x=i, j. Also, $t(v)$, $v \in V(G)$ is defined in the information routing section below. The estimation of channel information

from the received signal is outside the scope of this paper. Several papers have discussed this issue, e.g., [11].

The channel condition is quantized and graded to several statuses at the RFID tag, denoted by the weight of the link. The channel quality quantization function may be based on the SNR of the channel, for example, as an inverse function of it. The channel quantization is also not a main concern for discussion in this paper, and the details are not presented here. Good channel quality is quantized to small link weight, while bad channel quality is quantized to large link weight. An upper limit value w_{max} for the link weight is set. If the link weight is larger than w_{max}, there is no link between the two nodes or the link between the two nodes is not usable.

3.2. Information Routing

The channel aware information routing in the RFID tag network is to find a shortest path with good quality channels from the cluster head RFID tag to the command center. Since the range of RFID tags is not very large, the length of each hop does not vary much. We model the length of the path as the number of hops from a node to the command center. Then routing problem then turns into an optimization problem as follows:

$$\min \text{cost} = \left(\sum_{i,j \in \{1,2,\ldots,N\}} w_{ij} x_{ij} + \sum_{i,j \in \{1,2,\ldots,N\}} x_{ij} \right), \text{ where } x_{ij} = \begin{cases} 0 \\ 1 \end{cases} \tag{6}$$

The first term of the equation denotes the channel condition of each hop. If the channel of the hop is good, then the weight w_i assigned to that link is small. The second term of the equation denotes the length of the path. Thus, Equation (6) is used to find a routing path with both good link quality and short length.

Equation (6) is equivalent to:

$$\min \text{cost} = \sum_{i,j \in \{1,2,\ldots,N\}} w_{ij} x_{ij} \tag{7}$$

since the two terms are independent of each other.

Equation (7) can be solved using Dijkstra's Algorithm [12]. Given a graph with nonnegative weights and a starting node, Dijkstra's algorithm finds the shortest path from the starting node to other nodes in the graph. Its basic procedure is:

Starting node: u, weights of edges w_{ij}, $i, j \in \{1, 2, \ldots, N\}$

Initialization: $S = \{u\}$, $t(u) = 0$, $t(z) = w_{uz}$ for $z \neq u$

Iteration: Step 1: select a node $v \notin S$ such that $t(v) = \min_{z \notin S} t(z)$; $S = S \cup \{v\}$

Step 2: for each edge vz with $z \notin S$, $t(z) = \min\{t(z), t(v) + w_{vz}\}$

Iteration continues until $S = V(G)$ or $t(z) = \infty$ for each $z \notin S$

Length of shortest path between nodes u, v, is $d(u, v) = t(v)$ for all v.

The stopping rule of the iteration in our algorithm is modified to v = destination or t(z) = ∞ for each z ∉ S.

As for routing through holes, many papers have discussed this issue, such as [13,14]. Some complete void handling techniques include Greedy Perimeter Stateless Routing (GPSR) [15], Distance Upgrading Algorithm (DUA) [16], etc. We did not expand it here.

We illustrate the operation of the routing path selection algorithm through two examples. Target and cluster member RFID tags are neglected since they are not involved in the information routing process based on the design. The same simulation parameters are assumed as before.

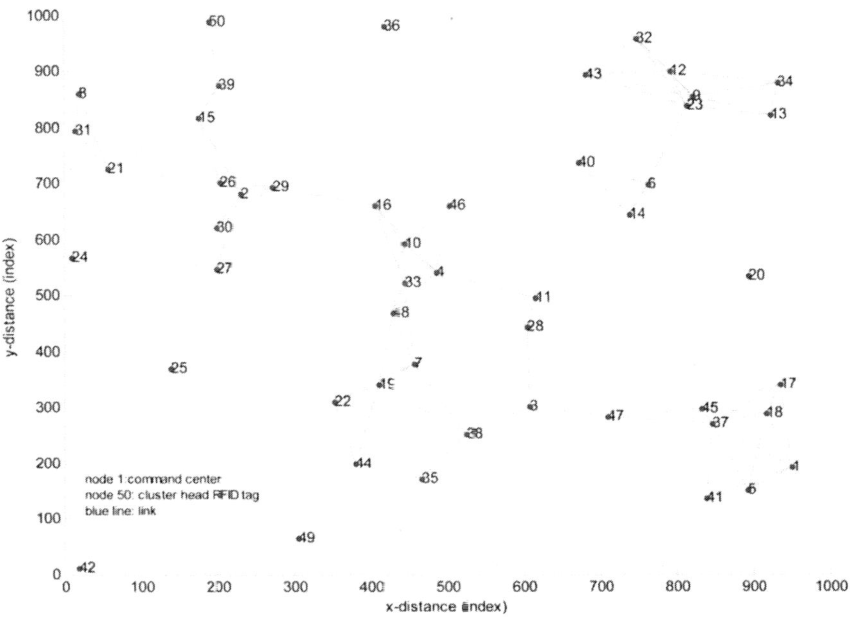

Figure 15. Topology of example 1.

In Example 1, 50 nodes are deployed. Each node represents an RFID tag. Node 1 represents the command center, Node 50 represents the cluster head RFID tag, and the blue line represents the link between two nodes. Weight of the link is quantized to 1 or ∞. The topology of example 1 is shown in Figure 15.

3.3. Simulation Results

The target's information routing path from node 50 to node 1 is depicted by the red line in Figure 16, which is the shortest path between node 50 and node 1.

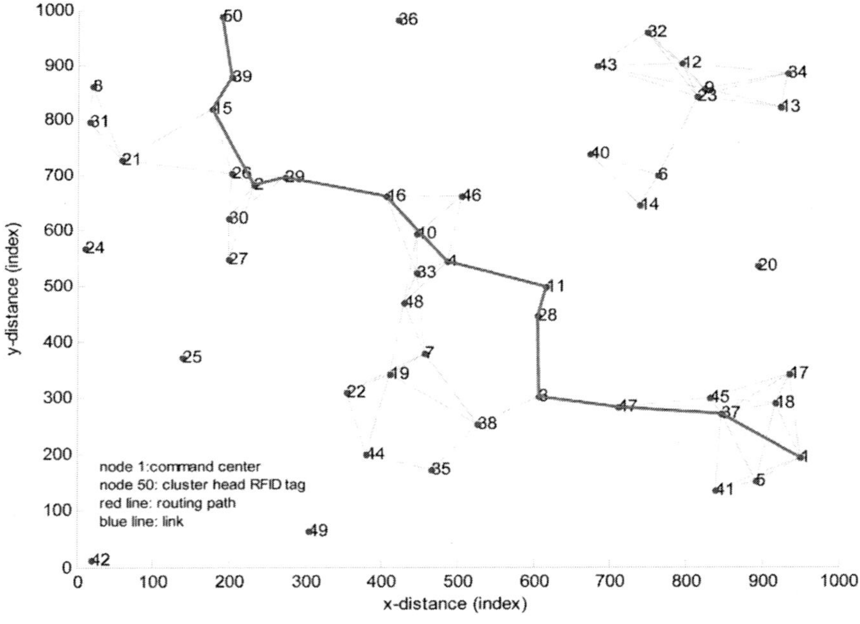

Figure 16. Routing path in example 1.

Nodes on the routing path are:

path = 50 39 15 2 29 16 4 11 28 3 47 37 1

and the cost of the path, which is the sum of weights of links on the path is:

$$\text{cost} = 12$$

Example 2 is a small network with ten nodes. Weights of the link are quantized to 5, 10, 20, and ∞. The weights of links are described by the matrix W.

$$W = \begin{bmatrix} index & 1 & 2 & 3 & 4 & 5 & 6 & 7 & 8 & 9 & 10 \\ 1 & 0 & 5 & \infty & \infty & \infty & \infty & \infty & 10 & 20 & \infty \\ 2 & 5 & 0 & 5 & \infty & \infty & \infty & \infty & \infty & \infty & \infty \\ 3 & \infty & 5 & 0 & 5 & \infty & \infty & 10 & \infty & \infty & \infty \\ 4 & \infty & \infty & 5 & 0 & 10 & \infty & \infty & \infty & \infty & \infty \\ 5 & \infty & \infty & \infty & 10 & 0 & \infty & \infty & \infty & \infty & 10 \\ 6 & \infty & \infty & \infty & \infty & \infty & 0 & \infty & \infty & \infty & \infty \\ 7 & \infty & \infty & 10 & \infty & \infty & \infty & 0 & \infty & 10 & 20 \\ 8 & 10 & \infty & \infty & \infty & \infty & \infty & \infty & 0 & \infty & \infty \\ 9 & 20 & \infty & \infty & \infty & \infty & \infty & 10 & \infty & 0 & \infty \\ 10 & \infty & \infty & \infty & \infty & 10 & \infty & 20 & \infty & \infty & 0 \end{bmatrix} \tag{8}$$

Node 1 represents the cluster head RFID tag, and Node 10 represents the command center. Simulation Results:

The routing path from node 1 to node 10 is:

$$path = \quad 1 \quad 2 \quad 3 \quad 4 \quad 5 \quad 10$$

and the cost of the path is:

$$cost = \quad 35.$$

4. HARDWARE IMPLEMENTATION CONSIDERATIONS

Three important considerations for hardware implementation of the proposed RFID tag structure are: (1) battery life; (2) antennas; and (3) information storage. These are discussed below.

The proposed RFID tags are active RF tags, meaning that they are designed to both receive commands and transmit coded information to other tags in the vicinity. Because an active RFID tag is powered by the internal energy, the lifetime of the tag is mainly dependent on the lifetime of the battery. However, they have built-in circuitry to turn on the transmitter only when they receive "wake up" commands. This extends battery life since higher current is drawn only when powering the transmit chain components. In the passive "sleep" mode, these tags act as simple RF receivers which do not require much battery power for functioning. The sleep mode enables the application to shut down the processor of unused modules, thereby saving power [17]. The receiver, which is kept active to react to an inquiry from the interrogator, therefore determines the shelf life of an active RFID tag. To further reduce the power of the RFID tag receiver two main technologies have been proposed: (1) a passive transceiver or burst switch allowing the tag to remain in a sleep mode until activated with RF energy; and (2) a smart buffer, which allows the controller to remain asleep while an incoming packet is buffered [18]. Use of RFID as the wake-up radio channel has been shown to provide a viable solution due to the low cost and ready off-the-shelf availability of RFID [19]. Furthermore, these circuits are able to wake up an entire neighborhood of nodes if a packet at a particular frequency is received.

These RFID tags must necessarily transmit their information in all directions to ensure that the information is assuredly picked up by other randomly distributed tags in the vicinity. This calls for a low-gain omnidirectional antenna, which is quite advantageous since such an antenna comes in smaller packages, and is therefore consistent with the small size of the tag. The antenna must be small enough to be attached to the tag, have omnidirectional or hemispherical coverage must provide maximum possible signal to the receiver, have a polarization matched to the enquiry signal regardless of the physical orientation of the tag, be robust and cheap [20]. Major considerations in antenna selection are antenna type, its impedance, RF performance when applied to the tag, and RF performance when the tag has other structures around it. Candidate omnidirectional antennas include the dipole and the folded dipole, with bandwidths of 10–15% and 15–20%, respectively [20]. Planar antennas are low cost, simple to manufacture, and have low profile suitable for RFID systems. The most common types of planar antennas for tags are folded dipoles, meander line antennas (MLAs) and spirals [21]. Planar elliptical patch antennas have been shown to be adequate for ultrawideband (UWB) applications in several bands. Such UWB antennas have been used in mobile handset devices with FR4 substrate using standard printed circuit board processes. The availability of high-contrast, low-loss ceramic materials permits significant antenna miniaturization, although they have higher loss characteristics [22]. However, their low profile and UWB operation make them quite attractive for use in RFID tags.

Other considerations include storage requirements. The storage size of the RFID tag depends on the number of possible targets to be monitored, size of the target's signal, size of the target ID, size of the tag ID, size of the keys, size of the counter, etc. Since these are application- and scenario-specific, it is difficult to assess the storage requirements in a general sense. It has been proposed that utilizing local storage of writeable RFID tags for inference and query processing makes the distributed approach a better solution with significantly reduced communication cost [23]. The query state primarily dominates the storage cost, and a larger numbers of queries may challenge the scalability of this approach. An approach to exploit the unique property of prime numbers to encode nodes in the path, and simultaneous congruence values to encode ordering between nodes in the path has been implemented and tested in [24]. The encoding scheme is based on the Fundamental Theorem of Arithmetic and the Chinese Remainder Theorem. Using the proposed path encoding scheme, it was shown possible to efficiently retrieve paths which satisfy the path condition in a query [24]. It is assumed in this paper that adequate storage size is available for proper functioning of the tag.

5. CONCLUSIONS AND FUTURE WORK

The design of a covert RFID tag network for target discovery and target's information routing is presented in this paper. The design and operations of the proposed algorithms are illustrated through examples. Simulation results clearly demonstrate the effectiveness of the design. In the design, we also considered the possible physical layer implementations, and considering that RFID tag's structure cannot be made very complex, we make the tradeoffs and do not incorporate too many advanced and accurate functionalities for the RFID tags. Although the initial RFID tag network design goals of the research have been achieved in this paper, further theoretical and experimental extensions are possible. For our future work, we plan to investigate more issues to make the design faultless and more practical, such as addressing the holes problem in the RFID tag network during information routing, data traffic and congestion. We will also consider the implementation of a small RFID tag network based on the design.

ACKNOWLEDGMENTS

This research was supported by US Air Force Office of Scientific Research (AFOSR) Grant # FA9550-06-1-0029. We appreciate valuable suggestions received from Jon Sjogren of AFOSR.

REFERENCES

1. Ngai, EWT; Moon, KL; Riggins, FJ; Yi, CY. RFID research: An academic literature review (1995–2005) and future research directions. Int. J. Prod. Econ **2008**, 112, 510–520.
2. Turner, L. The Evolution of Featureless Waveforms for LPI Communications. Proceedings of the IEEE National Aerospace and Electronics Conference (NAECON'91), Dayton, OH, USA, 20–24 May 1991; pp. 1325–1331.
3. Castelluccia, C; Avoine, G. Noisy Tags: A Pretty Good Key Exchange Protocol for RFID Tags. In Smart Card Research and Advanced Applications; Domingo-Ferrer, J, Posegga, J, Schreckling, D, Eds.; Springer: Berlin, Germany, 2006; Volume 3928, pp. 289–299.
4. Hancke, G. Modulating a noisy carrier signal for eavesdropping-resistant HF RFID. Elektrotech. Informationstech **2007**, 124, 404–408.
5. Predd, JB; Kulkarni, SR; Poor, HV. Distributed Learning in Wireless Sensor Networks. In Wireless Sensor Networks: Signal Processing and Communications Perspectives; Swami, A, Zhao, Q, Hong, Y-W, Tong, L, Eds.; John Wiley & Sons: Chichester, UK, 2007; pp. 185–214.
6. Chu, K-T; Wen, C-Y; Ouyang, Y-C; Sethares, WA. Adaptive Distributed Topology Control for Wireless Ad-Hoc Sensor Networks. Proceedings of 2007 International Conference on Sensor Technologies and Applications (SENSORCOMM 2007), Valencia, Spain, 14–20 October 2007; pp. 378–386.
7. Liu, J; Zhao, F; Petrovic, D. Information-directed routing in ad hoc sensor networks. IEEE J. Sel. Area. Comm **2005**, 23, 851–861.
8. Zhang, C. Cluster-based routing algorithms using spatial data correlation for wireless sensor networks. J. Commun **2010**, 5, 232–238. [Google Scholar]
9. Ok, C; Lee, S; Mitra, P; Kumara, S. Distributed routing in wireless sensor networks using energy welfare metric. Inform. Sci **2010**, 180, 1656–1670.
10. Hong, SH; Zhang, H; Song, IC; Chang, KH; Shin, D-B; Lee, H-S. ISO/IEC 18000-7 Based on RFID Multi-Hop Relay System. Proceedings of the 9th International Symposium on Communications and Information Technology (ISCIT 2009), Incheon, Korea, 28–30 September 2009; pp. 1450–1454.
11. Wada, T; Jamalipour, A; Okada, H; Ohuchi, K; Saito, M. Performance of Channel Information Estimation Method Utilizing Parity Check Bits for Turbo Coded Multi-Route Multi-Hop Networks.

Proceedings of the 2006 IEEE International Conference on Communications (ICC'06), Istanbul, Turkey, 11–15 June 2006; pp. 3688–3692.

12. West, DB. Introduction to Graph Theory, 2nd ed; Prentice Hall: Upper Saddle River, NJ, USA, 2001; pp. 97–98.
13. Ahmed, N; Kanhere, SS; Jha, S. The holes problem in wireless sensor networks: A survey. SIGMOBILE Mob. Comput. Comm. Rev **2005**, 9, 4–18.
14. Joshi, GP; Kim, SW. A distributed geo-routing algorithm for wireless sensor networks. Sensors **2009**, 9, 4083–4103.
15. Hightower, J; Borriello, G. Location system for ubiquitous computing. Computer **2001**, 34, 57–66.
16. Chen, S; Fan, G; Cui, J-H. Avoid 'void' in geographic routing for data aggregation in sensor networks. Int. J. Ad Hoc Ubiquitous Comput **2006**, 1, 169–178.
17. Cho, H; Baek, Y. Design and Implementation of an Active RFID System Platform. Proceedings of the International Symposium on Applications and the Internet Workshops (SAINTW'06), Phoeniz, AZ, USA, 23–27 January 2006.
18. Jones, AK; Hoare, RR; Dontharaju, SR; Tung, S; Sprang, R; Fazekas, J; Cain, JT; Mickle, MH. An Automated, Reconfigurable, Low-Power RFID Tag. Proceedings of the 43rd Annual Design Automation Conference (DAC'06), San Francisco CA, USA, 24–28 July 2006; pp. 131–136.
19. Jurdak, R; Ruzzelli, AG; O'Hare, GMP. Multi-Hop RFID Wake-Up Radio: Design, Evaluation and Energy Tradeoffs. Proceedings of 17th International Conference on Computer Communications and Networks (ICCCN'08), St Thomas, US Virgin Islands, USA, 3–7 August 2008.
20. Foster, PR; Burberry, RA. Antenna Problems in RFID Systems. Proceedings of the IEE Colloquium on RFID Technology, London, UK, 25 October 1999; pp. 3/1–3/5.
21. Serrano, R; Blanch, S; Jofre, L. Small Antenna Fundamentals and Technologies: Future Trends. Proceedings of the 1st European Conference on Antennas and Propagation (EuCAP 2006), Nice, France, 6–10 November 2006.
22. Ren, Y-J; Lai, C-P; Chen, P-H; Narayanan, RM. Compact ultrawideband UHF array antenna for through-wall radar applications. IEEE Antennas Wirel. Propag. Lett **2009**, 8, 1302–1305.
23. Cao, Z; Diao, Y; Shenoy, P. Architectural Considerations for Distributed RFID Tracking and Monitoring. Proceedings of the ACM Workshop on Networking Meets Databases (NetDB), Big Sky, MT,

USA, 14 October 2009; Available online http://lass.cs.umass.edu/papers/pdf/netdb09.pdf (accessed on 15 September 2011).

24. Lee, C-H; Chung, C-W. Efficient Storage Scheme and Query Processing for Supply Chain Management Using RFID. Proceedings of the 2008 ACM SIGMOD International Conference on Management of Data (SIGMOD'08), Vancouver, BC, Canada, 9–12 June 2008; pp. 291–302.

INDEX